Laboratory Experiments Teacher's Edition

This book was printed with acid-free recycled content paper, containing 10% POSTCONSUMER WASTE.

HOLT, RINEHART AND WINSTON

A Harcourt Classroom Education Company

Austin • New York • Orlando • Atlanta • San Francisco • Boston • Dallas • Toronto • London

Holt Science Spectrum: A Physical Approach

Laboratory Experiments Teacher's Edition

Author
E. David Thielk
Science Educator
Peninsula College
Port Townsend, WA

Reviewers
Rosa Hemphill
Science Teacher
High School
California

Bert Sherwood
Science Instructional Specialist
Socorro I.S.D.
El Paso, TX

Richard Sorensen
Vernier Software and Technology
Beaverton, OR

Safety Reviewer
Gregory Puskar
Laboratory Manager, Physics Department
West Virginia University
Morgantown, WV

Cover: basketball image: David Madison/Tony Stone Images; celestial background image: Corbis Images.

Printed in the United States of America

ISBN 0-03-055582-5

4 5 6 862 03 02 01

Teacher's Edition Contents

Using the CBL™ System

Integrating Calculator-Based Laboratory™ (CBL), probes, sensors, and graphing calculators into your laboratory

The procedures for the experiments featured in this laboratory manual are written for the Texas Instruments Calculator-Based Laboratory (CBL) System. Most of the experiments also require use of the physical science (PHYSCI) program by Vernier Software and Technology. In each experiment, the step-by-step instructions refer specifically to the menus that will appear on your students' calculator screens as they are using this program.

Texas Instruments offers a variety of graphing calculators that can be used with the CBL System. A list of these calculators is shown below.

- TI-73
- TI-82
- TI-83
- TI-83 Plus
- TI-86
- TI-89
- TI-92
- TI-92 Plus

Although the instructions given in the experiments were written to work for all TI graphing calculators, some minor adaptations may need to be made depending on which calculators your students use.

Getting Started

System requirements

You will need TI-GRAPH LINK™ software and a TI-GRAPH LINK computer-interface cable to get started using the PHYSCI program with your TI graphing calculator and CBL. These are available from CENCO, Vernier Software and Technology, or educational- and office-supply companies.

internetconnect

TOPIC: Free Vernier Software for the CBL™
GO TO: go.hrw.com
KEYWORD: HK1 VERNIER

Visit the go.hrw.com Web site for the version of Vernier's PHYSCI lab software that will work best with *Holt Science Spectrum* and the graphing calculators your students will use. At this site, you can download the lab software from the web onto your computer's hard drive. Instructions on how to get it done are found at the Web site.

After you download the software, open and read the text file. It contains more detailed information about the PHYSCI program by Vernier Software and Technology. It might be helpful to print a copy of this file so you can refer to it whenever you need to.

You may also want to refer to the *CBL Made Easy!* booklet that comes with your purchase of a CBL. The most up-to-date version of the text file and this booklet can be found on the Vernier web site, www.vernier.com.

Downloading the lab software to a calculator

1. To download the PHYSCI group file from your computer to your calculator, follow the directions in the manual that comes with the TI-GRAPH LINK cable. **Note:** Because of the size of the PHYSCI program, it is recommended that all other programs be removed before loading the PHYSCI program onto a calculator.
2. When the transfer is complete, a set of programs will appear on the calculator's screen. The set will include the program titled PHYSCI and a list of related PHYSCI programs that are to be used by the main PHYSCI program.
3. Use the link-to-link cable to copy the programs onto each student's calculator.
4. Students should keep all of the PHYSCI programs on their calculators so that they can use them throughout the year. Because these programs are locked, students will not be able to change the programs on their own calculators.

Using the Software

Recognizing the software's flexibility

To use the PHYSCI program, simply follow the instructions given in each experiment's procedure. The PHYSCI program can calibrate sensors and probes, collect data, and perform data analysis. This enables you to use the CBL to its full potential without writing your own programs or using many different interfaces.

The PHYSCI program is designed to be very flexible. There are a variety of data-collection modes and probes/sensors available. You even have the ability to use more than one probe or sensor at a time so that several different measurements can be made simultaneously.

Contacts for More Information or Technical Support

TEXAS INSTRUMENTS
http://www.ti.com
technical support: 1-800-TI-CARES
programming assistance: (972) 917-8324
E-mail: ti-cares@ti.com
address: Texas Instruments
P.O. Box 650311, MS 3962
Dallas, TX 75265

VERNIER SOFTWARE AND TECHNOLOGY
http://www.vernier.com
technical support: (503) 277-2299
E-mail: info@vernier.com
address: Vernier Software and Technology
13979 SW Millikan Way
Beaverton, OR 97005-2886

MATERIALS LIST

The following list indicates the quantities needed for 10 lab groups to perform the labs in this lab manual. If you have a different number of lab groups to plan for, you will need to adjust the quantities accordingly.

WARD'S is the supplier for *Holt Science Spectrum.* WARD'S catalog numbers are provided for all the materials you will need to do these labs except for the probes, sensors, and CBL equipment, which you can order from either Vernier Software and Technology or Texas Instruments.

For Chemicals and Equipment:

WARD'S Natural Science Establishment
5100 W. Henrietta Road
P.O. Box 92912
Rochester, NY 14692-9012

For Probes, Sensors, and CBL Equipment:

VERNIER SOFTWARE AND TECHNOLOGY
http://www.vernier.com
13979 SW Millikan Way
Beaverton, OR 97005-2886
(503) 277-2299
E-mail info@vernier.com

or

TEXAS INSTRUMENTS
http://www.ti.com
P.O. Box 650311, MS 3962
Dallas, TX 75265
1-800-TI-CARES
E-mail ti-cares@ti.com

Probes, Sensors, and CBL Equipment

Qty.	Description	Used in lab(s)
10	CBL	throughout
10	TI graphing calculator and black link cable	throughout
10	Conductivity probe (and adapter cable)	4, 6
10	Current probe (and adapter cable)	13
10	Dual channel amplifier	13
10	Dual-range or student force sensor (and adapter cable)	2, 8, 9, 16
10	Light sensor	12, 15
10	Magnetic field sensor (and adapter cable)	14
10	Microphone (CBL-compatible)	11
10	Motion detector	1
10	pH electrode (and adapter cable)	19
20	Temperature probe	5, 10, 17, 18
10	Voltage probe (and adapter cable)	13

Chemicals and Consumable Materials

Qty.	Description	Used in lab(s)	WARD'S no.
2	Alcohol, rubbing, 16 oz bottle	4	39 R 4915
6	Aprons, disposable, polyethylene, 100/box	throughout	15 R 1050
2	Baking soda, 1 lb	4	37 R 5467
1	Cement, duco, 1 fl oz	2, 4, 5, 6, 10	15 R 1100
2	Detergent, household, liquid, 14 oz	4	37 R 2268
2	Epsom salt, 16 oz	4	37 R 2860
1	Food coloring, 4 bottles/pkg.	17	15 R 0071
12	Gloves, disposable, medium, 100/pkg.	throughout	15 R 1071
1	Hydrochloric acid, 1.0 M, 1.0 L	5	37 R 6563
1	Paper towels, 1 roll	10	15 R 9844
1	Plant food, 1.5 lb	4	20 R 6020
1	Plastic spoons, 100/pkg.	2	15 R 9800
1	Potassium phosphate, monobasic	19	37 R 4845
1	Salt, iodized, 26 oz	4	37 R 5482
8	Sand, fine	2, 18	20 R 7423
1	Sodium chloride, crystals, reagent	6	37 R 5480
10	Sodium chloride for 2% solution	13	37 R 9553
1	Sodium thiosulfate	5	37 R 5692
1	Sugar, granular, 5 lb bag	4	39 R 3180
1	Sulfuric acid, 1.0 M, 1.0 L	19	37 R 8618
1	Tape, masking, 3/4 in. × 60 yd	14	15 R 9828
3	Water, distilled, 1 gal	4, 5, 6, 13, 19	88 R 7005
1	Wipes, disposable, 280/box	4, 6	15 R 1025

Equipment and Reusable Materials

Qty.	Description	Used in lab(s)	WARD'S no.
60	Alligator connector clips, 2 in., screw	13	15 R 9473
10	Aquarium, 10 gal	17	21 R 5241
5	Balance, triple beam	1, 10	15 R 6057
10	Bathroom scale, lb/kg	8	15 R 3800
10	Battery, 6 V	13	15 R 3263
20	Battery, size D, 1.5 V alkaline	15	15 R 3247
30	Beaker, 100 mL	4, 6, 19	17 R 4020
20	Beaker, 150 mL	18	17 R 4030
20	Beaker, 50 mL	5	17 R 4010
10	Beaker, 600 mL	4, 6, 19	17 R 4060
20	Clamp holder	18	15 R 0646
10	Clamp holder, adjustable	12, 18	15 R 0770
10	Dropping bottle, flint glass, 15 mL	19	17 R 6018
10	Flashlight	15	15 R 3264

Equipment and Reusable Materials, continued

Qty.	Description	Used in lab(s)	WARD'S no.
10	Glass plate, 4 in. × 4 in.	12	15 R 3812
1	Glass stirring rods, 10/pkg.	17	17 R 6005
10	Gloves, heat defier	17	15 R 1095
30	Goggles, general purpose	throughout	15 R 3046
20	Graduated cylinder, polypropylene, 10 mL	5	18 R 1705
10	Graduated cylinder, polypropylene, 100 mL	4	18 R 1730
10	Graduated cylinder, polypropylene, 2000 mL	2	18 R 1765
10	Hall's carriage	9	15 R 8500
10	Hot plate	10,17	15 R 8010
10	Jar cap, 53 mm	2, 17	17 R 2133
10	Jar cap, 58 mm	2	17 R 2137
10	Jar, clear polystyrene, 2 oz, 53 mm	2, 17	18 R 1632
10	Jar, clear polystyrene, 4 oz, 58 mm	2	18 R 1633
10	Lamp, clamp with reflector	18	36 R 4168
10	Lamp, desktop, incandescent	12	15 R 5036
10	Light bulb, 150 W, clear	18	36 R 4173
10	Magnet, steel bar	14	13 R 0115
10	Marker, black	16	15 R 3083
10	Meterstick, wood	1, 9, 14, 15, 16	15 R 4065
1	Nails, 8d, common, 36/pkg.	13	15 R 9478
1	Pencils, No. 2, 12/box	14	15 R 9816
1	Rope (clothesline), 5/32 in. × 96 ft	8	15 R 3991
10	Ruler, 12 in., standard/metric	13, 18	15 R 4650
20	Stopper, black rubber, 1 hole, size 3	10	15 R 8483
10	Stopwatch, digital	5, 10, 16, 17	15 R 0512
1	String, 1/2 lb spool	1, 2, 8, 9,	15 R 9863
10	Support base, rectangular	1, 6, 12, 18	15 R 0719
10	Support ring, nickel plated	1	15 R 0707
10	Support rod, 24 in.	18	15 R 0663
10	Support rod and clamp	6, 8	15 R 0696
10	Test-tube rack, 16 mm tubes	5	18 R 4213
20	Test tube with rim, Pyrex®, 15 mm × 125 mm	5	17 R 0620
10	Thermometer, red alcohol, −20°C to 110°C (if temperature probes are not available)	5, 10, 17, 18	15 R 1462
10	Tongs, 8 in.	17	14 R 0960
10	Tray tote box, polyethylene	2	18 R 9881
10	Tray with cover, polyethylene, 1 gal	13	18 R 3650
10	Wash bottle, wide mouth, 1000 mL	4, 5, 6	18 R 4117
10	Wash bottle, wide mouth, 500 mL	19	18 R 4116
10	Water bath, 35°C–65°C	5	14 R 5759
10	Weight set, 10 mg–500 mg	1	15 R 3760
1	Wire, copper, bare, magnet, 1 lb spool	2	15 R 9235
1	Wire, copper, PVC coated, 100 ft	13	16 R 0549

Locally Obtained Materials

These materials are not carried by WARD'S but should be easy to find.

Qty.	Description	Used in lab
10	Bicycle, multi-gear	8
10	Board, at least three times as long as a chair's height	9
30	Books, heavy	9
20	Bottle, plastic beverage	10
10	Cardboard, piece	15
10	Chair	9
1	Computer with a printer	15
1	Drill to make a hole in jar's lid	17
10 m	Fishing line, 25 lb	16
20 oz	Milk of magnesia	4
10	Object with a mass of 2–6 kg	8
10	Snap, swivel	16
10	Sock, cotton	10
10	Sock, wool	10
10	Stringed musical instrument	11
30–45	Sunglasses, various, some polarized	12
10 ft	Teflon® tape	17
10	Tube, small plastic	16
1.5 L	Water samples from 2–3 different sources	19

Designing a Pendulum Clock

MATERIALS (for each lab group)

- ✓ motion detector
- ✓ CBL
- ✓ TI graphing calculator and black link cable
- ✓ several pieces of string, each of a different length
- ✓ several different hooked masses ranging from 5g–25g
- ✓ ring stand or table support
- ✓ metric ruler
- ✓ balance
- ✓ graph paper (optional)

Students should wear safety goggles and gloves while performing this experiment.

If students are working in the laboratory, they should also wear laboratory aprons.

Ensure that students set up their pendulums so that the pendulums are able to swing safely without hitting anyone.

Remind students that they should only swing the pendulums when they are collecting data.

Make sure that students wash their hands when they are done with the experiment.

▶ Objectives

In this lab, students will

- ***make hypotheses*** explaining how they think changing one variable of a pendulum will affect how the pendulum swings.
- ***conduct*** experiments to test their hypotheses.
- ***evaluate*** the precision and accuracy of a pendulum for measuring time.

▶ Planning for the Experiment

Organizing your class

Having two to four students in each lab group works best.

Recommended time

1–2 lab periods

Preparing the CBL System

- If possible, have a motion detector, a CBL, and a graphing calculator connected for each group ahead of time. The calculators should be loaded with the Vernier PHYSCI program.
- The pendulum should be set up so that it can swing parallel to the table support and the motion detector. The pendulum shown on page 2 is set up to swing into and out of the page.
- The motion detector should be set up 0.5 m away from the pendulum and away from sound sources like air vents that may interfere with its readings.
- Remind students to make sure that there are no stationary objects between the motion detector and the pendulum.

Preparing other materials

- You'll need to have several pieces of string of different lengths handy for the groups that choose to vary the length of the pendulum string.
- For the groups that choose to vary the mass attached to the string, you'll need to have a range of hooked masses available. Have students use a balance to measure the masses.
- Make sure that the ring stand or table support can support the largest mass that is available to students.

▶ Procedural Notes

General

- Do not allow students to use excessive displacements when pulling the mass back to start the pendulum swinging.
- Emphasize to students the importance of treating the pendulum exactly the same way in each of the three trials.

- You may want to have more advanced students compare their data with the expected period of the pendulum by using the following equation:

$$\text{period} = 2\pi \times \sqrt{\frac{\text{length of string (m)}}{9.8\ \text{m/s}^2}}$$

Specifics

- **Step 7** Make sure that students attach the mass securely to the string and the string securely to the table support.
- **Step 12** Remind students that when a hypothesis is made, it doesn't matter whether the hypothesis is correct or not. Emphasize to them that being able to disprove a hypothesis is just as important as being able to prove one.

▶ Sample Data

- Groups that vary the length of the string will see the greatest change in the period of the pendulum.
- Groups that vary the distance they pull the pendulum back may see slight changes in the period.
- Groups that vary the mass attached to the string should not see any change in the period.
- This sample data was collected to test the following hypothesis: Increasing the length of the string will increase the period of the pendulum.

Pendulum Data

Length of string (cm)	Period of the pendulum (s)			
	Trial 1	Trial 2	Trial 3	Average
10	0.70	0.80	0.80	0.77
20	1.20	0.90	0.90	1.00
30	1.30	1.20	1.10	1.20

Answers

▶ Analyzing Your Results

1. Answers will vary. For the sample data provided:

$$\text{average period 1} = \frac{0.70\ \text{s} + 0.80\ \text{s} + 0.80\ \text{s}}{3} = 0.77\ \text{s}$$

$$\text{average period 2} = \frac{1.20\ \text{s} + 0.90\ \text{s} + 0.90\ \text{s}}{3} = 1.00\ \text{s}$$

$$\text{average period 3} = \frac{1.30\ \text{s} + 1.20\ \text{s} + 1.10\ \text{s}}{3} = 1.20\ \text{s}$$

2. Graphs will vary. For the hypothesis and sample data provided:

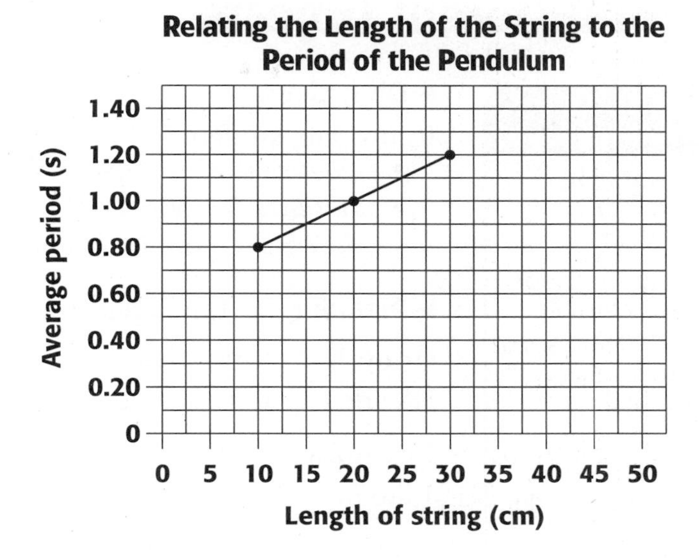

▶ Reaching Conclusions

3. Some students will disprove their hypothesis, while others will prove their hypothesis. For the sample data provided, the graph clearly shows that the longer the string, the greater the period of the pendulum.

4. No, because there is too much uncertainty in the exact length of the string and the mass of the object attached to it to measure time that precisely.

▶ Defending Your Conclusions

5. No; results may be precise without being accurate. For example, if the motion detector were not working properly, it may make similar measurements for each trial, but these measurements may not reflect what is really happening.

▶ Expanding Your Knowledge

1. Some devices students can research are Christian Huygen's pendulum clock, John Harrison's chronometer, and the modern quartz and atomic clocks. Students should find that modern clocks are much more accurate and precise than these older devices.

2. Provide students with some guidelines as they design their clocks. The biggest challenge for them will be finding a way to keep the pendulum swinging for a full minute without affecting the period.

Comparing the Buoyancy of Different Objects

MATERIALS (for each lab group)

- ✓ force sensor (and adapter cable)
- ✓ CBL
- ✓ TI graphing calculator and black link cable
- ✓ 2 jars of different sizes, each with a small wire loop attached
- ✓ large container or laboratory sink
- ✓ 2 L graduated cylinder
- ✓ tap water
- ✓ sand
- ✓ piece of sturdy string
- ✓ scoop
- ✓ graph paper (optional)

Students should wear safety goggles and gloves while performing this experiment.

If students are working in the laboratory, they should also wear laboratory aprons.

Make sure students handle glassware with caution and dispose of any broken glass in a specially designated container.

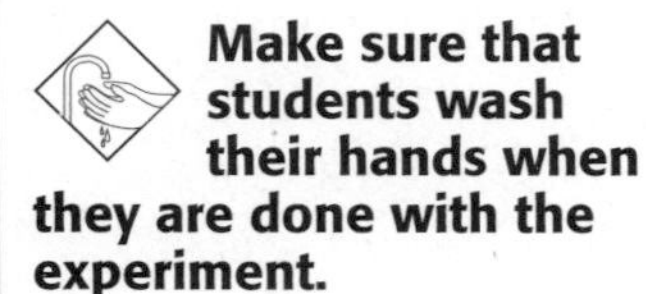

Make sure that students wash their hands when they are done with the experiment.

▶ Objectives

In this lab, students will

- ***measure*** the force exerted on an object in air and underwater.
- ***calculate*** the buoyant force water exerts on the object.
- ***relate*** the volume and mass of the object to its buoyancy.

▶ Planning for the Experiment

Organizing your class

Having two to four students in each lab group works best.

Recommended time

1 lab period

Preparing the CBL System

- If possible, have a force sensor, a CBL, and a graphing calculator connected for each group ahead of time. The calculators should be loaded with the Vernier PHYSCI program.
- This lab's procedure is written for a dual-range force sensor, which can measure forces up to 50 N. Student force sensors can't measure forces greater than 10 N. If student force sensors are used, the CBL System will need to be set up differently, as described in the *Procedural Notes* section.

Preparing other materials

- If possible, supply glass jars of many different sizes. (Glass jars require less sand to sink than plastic jars do.)
- Securely attach a wire with a small loop around the neck of each jar so the jars can be easily suspended in air or water.
- Make sure that there are no sharp edges on the wires wrapped around the jars or on the strings the jars will be suspended from. Make sure that the strings the jars will be suspended from are strong enough to support the largest jar filled with sand.

▶ Procedural Notes

General

- Provide paper-lined trays for students to place wet jars on and to catch spills. Remind students to be careful when handling wet glass because it can be very slippery. Be sure to clean up any spills immediately, especially if they are on the floor.
- For larger jars, the switch on the force sensor may need to be set to 50 N. Students will need to go back to the MAIN MENU and adjust the settings before collecting data.

NOTE: It may be difficult to get a reading of 0 on the force sensor. Have students adjust the sensor to get as close to 0 as possible.

Specifics

- **Step 2** You may want to assemble a ring stand with a large clamp to help stabilize the graduated cylinder.
- **Steps 5 and 6** If students are using student force sensors, they should substitute the following steps for steps 5 and 6:
  5. Go to the MAIN MENU, and select SET UP PROBES. Enter "1" as the number of probes. Select FORCE from the SELECT PROBE menu. Enter "1" as the channel number.
  6. Select USE STORED from the CALIBRATION menu. Then select STUDENT FORCE from the FORCE PROBE menu.
- **Step 12** You may want to have students do this step over a padded cardboard box in case the jar breaks loose.

▶ Sample Data

Table 1 Jar Volumes

Jar	Volume of water (mL)	Volume of water and jar (mL)	Volume of jar (mL)
1	785	955	170
2	740	977	237

Table 2 Forces Acting on the Jars

Jar	Trial	Force acting on jar in air (N)	Force acting on jar underwater (N)	Buoyant force acting on jar (N)	Average buoyant force (N)
1	1	3.01	1.29	1.72	1.70
	2	2.19	0.50	1.68	
	3	1.62	– 0.08	1.70	
2	1	3.67	1.30	2.37	2.37
	2	2.90	0.53	2.37	
	3	2.30	– 0.08	2.38	

Note: The equation students use to calculate the buoyant force acting on a jar is written to take direction into account so that the buoyant force will be positive.

Answers

▶ Analyzing Your Results

1. Answers will vary. For the sample data provided:

volume of jar 1 = 955 mL – 785 mL = 170 mL
volume of jar 2 = 977 mL – 740 mL = 237 mL

2. Answers will vary. The calculation for jar 1 for the first trial is shown below. Answers for all three trials for both jars are listed in **Table 2.**

$$\textbf{buoyant force acting on jar 1} = 3.01\ \text{N} - 1.29\ \text{N} = 1.72\ \text{N}$$

3. Answers will vary. For the sample data provided:

$$\textbf{average buoyant force for jar 1} = \frac{1.72\ \text{N} + 1.68\ \text{N} + 1.70\ \text{N}}{3} = 1.70\ \text{N}$$

$$\textbf{average buoyant force for jar 2} = \frac{2.37\ \text{N} + 2.37\ \text{N} + 2.38\ \text{N}}{3} = 2.37\ \text{N}$$

4. Students should exchange data with another lab group that tested different-sized jars. Graphs will vary. For the sample data provided:

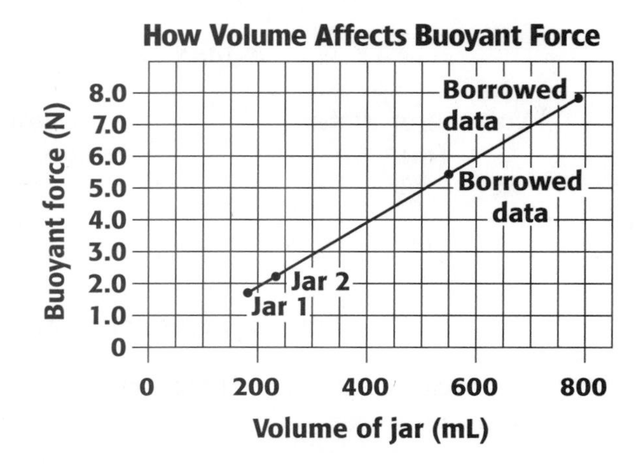

▶ Reaching Conclusions

5. The greater the volume of an object, the greater the buoyant force acting on it. The buoyant force was not affected when mass was subtracted from each jar.

6. Answers will vary. Based on the graph of the sample data, a float with a volume of 500 mL would have a buoyant force of 5.0 N.

▶ Defending Your Conclusions

7. For the boat to float, the buoyant force acting on it must be greater than the force of gravity acting on it.

▶ Expanding Your Knowledge

1. Provide each student with a single square foot of thick aluminum foil, and have them design their own boats. Then hold a contest to see whose boat can carry the most pennies before sinking. Have students explain why the winner's boat didn't sink.

Predicting the Physical and Chemical Properties of Elements

MATERIALS (for each student)

✓ TI graphing calculator (or graphing software or graph paper and a pencil)

Students should wear safety goggles while performing this experiment.

If students are working in the laboratory, they should also wear gloves and laboratory aprons.

Make sure that students wash their hands when they are done with the experiment.

▶ Objectives

In this lab, students will

- ***make graphs*** for some of the physical and chemical properties of the first 20 elements in the periodic table.
- ***observe patterns*** in the properties of elements based on their positions in the periodic table.
- ***predict*** the properties of two elements.

▶ Planning for the Experiment

Organizing your class

Having four students in each lab group works best.

Recommended time

1 lab period

Preparing materials

- This lab's procedure is written for a TI graphing calculator, but any graphing calculator or graphing software will work.
- If you prefer, students can use graph paper. Another option is for students to combine two or more graphs on one calculator. To do this, students must enter the data in separate lists in the STAT menu. Different graphs can be displayed by changing the Ylist in the STAT PLOT menu. This option is only recommended for students that are already familiar with the graphing capabilities of the calculator.

▶ Procedural Notes

General

- Be sure that students use the (–) button (and not the subtraction button) when entering negative values.
- When the groups are asked to compare the properties of elements, it will be helpful if they put their calculators together so that all four screens can be viewed at the same time.
- The trends in the periodic table and the graphs that are generated are general. Remind students that there are some exceptions to these general trends.

Specifics

- **Step 1** If students will make more than one graph per calculator, they should delete all six lists instead of just three. Lists L4–L6 can be deleted the same way as lists L1–L3.
- **Step 3** Students must name the lists, or columns, they will use to enter data. These lists may already be named "L1," "L2," etc.

- **Step 4** Make sure that students do not enter the numbers 3 and 9 in the first column. These are the atomic numbers of the elements missing from the table.
- **Step 5** When entering negative data, students will have to enter the negative sign, (–), first and then the number.
- **Step 6** Be sure that students use the line graph option when configuring the STAT PLOT menus. Using a scattergram option will not allow them to see patterns as easily.
- **Step 7** Students must label the Xlist and Ylist with the exact same names that were used in step 3. Otherwise, the calculator will indicate that there is an error.

▶ Sample Data

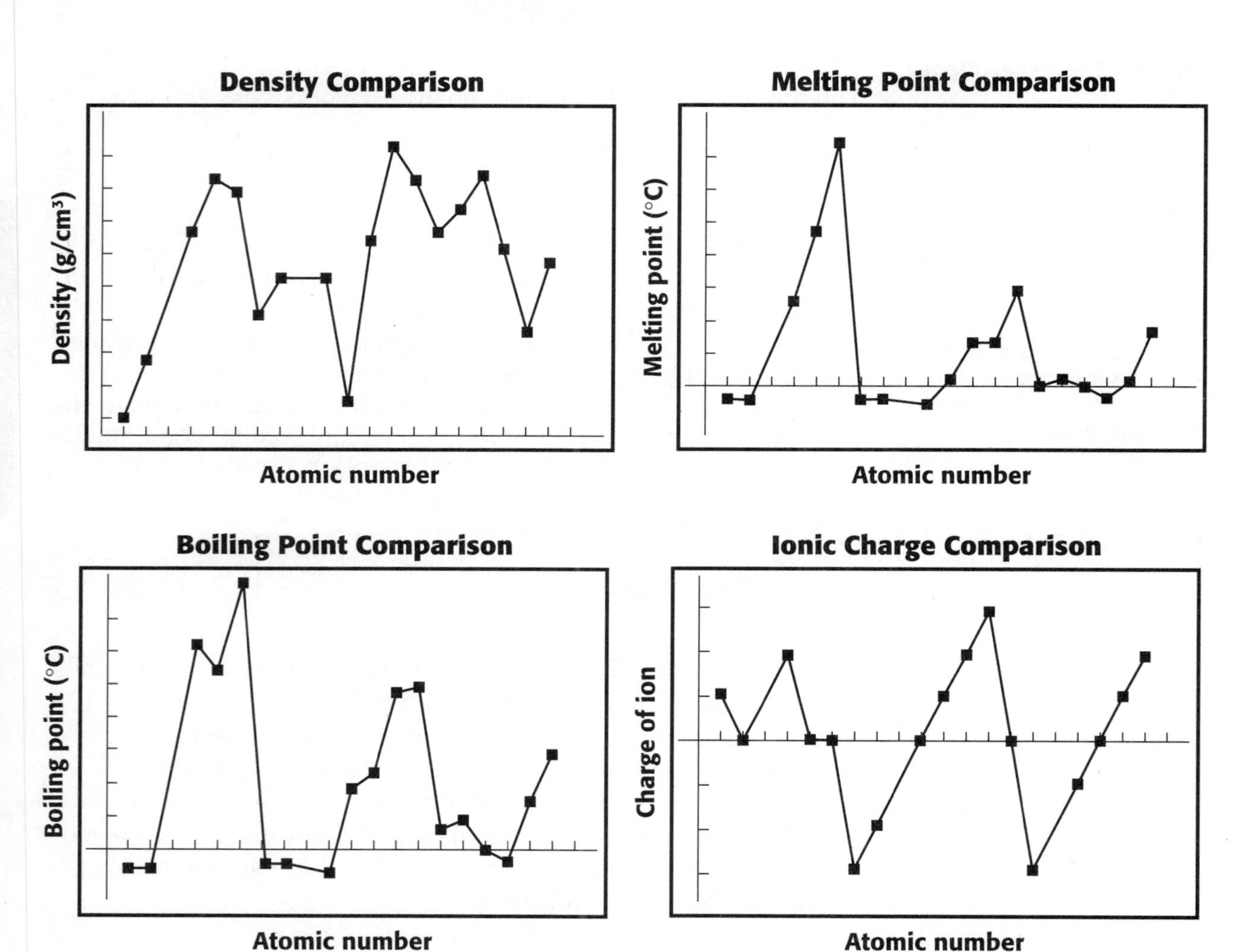

Actual Values

Element	Atomic number	Density (g/cm³)	Melting point (°C)	Boiling point (°C)	Charge of ion
Li	3	0.53	186	1326	1+
F	9	1.50	– 220	– 188	1–

Answers

▶ Analyzing Your Results

1. Estimated values will vary. Answers should be reasonably close to the actual values listed in the table on the previous page. Lithium is an alkali metal, or Group 1, element.
2. Estimated values will vary. Answers should be reasonably close to the actual values listed in the table on the previous page. Fluorine is a halogen, or Group 17, element.

▶ Reaching Conclusions

3. Carbon and silicon belong to the same family—Group 14. The graphs of the physical properties (density, melting point, and boiling point) show that both of these elements are on or near a peak. The graph of the chemical property (charge of ion) shows that neither atom forms ions.
4. Nitrogen and phosphorus also belong to the same family—Group 15. The graphs of the physical properties show that both of these elements are in a trough. The graph of the chemical property shows that atoms of nitrogen form ions with a 3− charge, just as phosphorus atoms do.

▶ Defending Your Conclusions

5. Answers will vary but might include the estimated values were close to the actual values in most cases. There might be some variation because values were estimated using the patterns of the graphs, which do not take exceptions into account.

▶ Expanding Your Knowledge

1. Either let students pick an element or assign them the more common or interesting elements. Have students spend a class period in the library or using the Internet to do their research. There are many interesting historical notes and uses of elements that students may discover and report.

Additional Notes

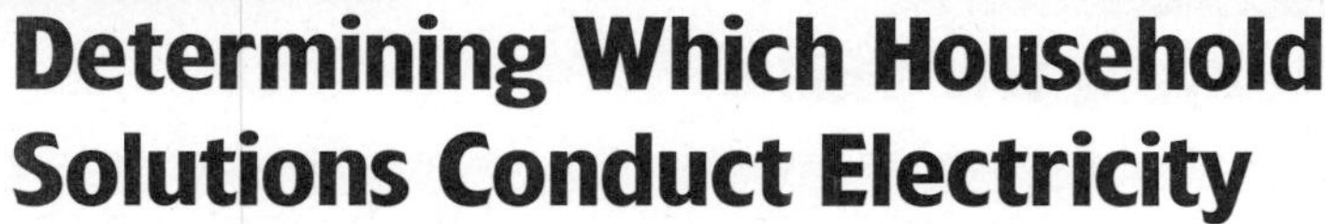

Determining Which Household Solutions Conduct Electricity

MATERIALS (for each lab group)

- ✓ conductivity probe (and adapter cable)
- ✓ CBL
- ✓ TI graphing calculator and black link cable
- ✓ table sugar solution
- ✓ table salt solution
- ✓ baking soda solution
- ✓ Epsom salt solution
- ✓ dishwashing detergent solution
- ✓ rubbing alcohol solution
- ✓ milk of magnesia solution
- ✓ chemical fertilizer solution
- ✓ 100 mL beakers (3)
- ✓ rinse bottle filled with deionized water
- ✓ large beaker
- ✓ tissue paper

▶ Objectives

In this lab, students will

- ***measure*** the conductivities of several dilute samples of household solutions.
- ***relate*** the conductivity of each sample to the kinds of bonds that are present in each substance.
- ***determine*** which solutions are likely to be contributing to the increased conductivity of a river.

▶ Planning for the Experiment

Organizing your class

Having four students in each lab group works best.

Recommended time

1 lab period

Preparing the CBL System

- The calculators should be loaded with the Vernier PHYSCI program.

Preparing other materials

- You can purchase 1000 μS NaCl solution to calibrate the conductivity probes if neccesary.
- Use the following information to prepare each solution:

Substance	Solution concentration (mass percent)	Preparation instructions
Table sugar	1% $C_{12}H_{22}O_{11}$	Dissolve 10 g of table sugar (sucrose) in enough deionized water to make 1 L.
Table salt	0.5% NaCl	Dissolve 5 g of table salt (sodium chloride) in enough deionized water to make 1 L.
Rubbing alcohol	1% C_3H_8O	Dilute 18 mL of 70 percent rubbing alcohol with enough deionized water to make 1 L.
Milk of magnesia	1% $Mg(OH)_2$	Dissolve 10 g of magnesium hydroxide in enough deionized water to make 1 L.
Epsom salt	1% $MgSO_4$	Dissolve 10 g of magnesium sulfate in enough deionized water to make 1 L.
Baking soda	1% $NaHCO_3$	Dissolve 10 g of sodium hydrogen carbonate in enough deionized water to make 1 L.
Dishwashing detergent	5% $C_{12}H_{25}SO_4Na$	Dilute 50 mL of sodium lauryl sulfate (or dishwashing detergent) with enough deionized water to make 1 L.
Chemical fertilizer	1% solution	Thoroughly mix 10 g of fertilizer with enough deionized water to make 1 L. Allow the mixture to stand for several hours before using.

Students should wear safety goggles, gloves, and laboratory aprons while performing this experiment.

Make sure you have Material Safety Data Sheets (MSDSs) available for all chemicals. If a chemical comes in contact with a student's bare skin, have the student rinse the affected area immediately and continuously with lukewarm water. Remind students that they should never eat or drink anything in the laboratory.

Make sure students handle glassware with caution and dispose of any broken glass in a specially designated container.

Rubbing alcohol is flammable, so make sure that no lit Bunsen burners or other heat sources are present in the laboratory.

Follow the disposal instructions given in the *Procedural Notes* section. Make sure that students wash their hands when they are done with the experiment.

▶ Procedural Notes

General

- Emphasize that it is very important for students to label their samples clearly.
- Make sure students use care when handling the conductivity probe. When testing a sample, the probe should not hit the sides or bottom of the beaker.
- Remind students to rinse the probe with deionized water between samples to avoid contamination.

Specifics

DISPOSAL INSTRUCTIONS

- *Table sugar* (sucrose), *table salt* (sodium chloride), *Epsom salt* (magnesium sulfate), *baking soda* (sodium hydrogen carbonate), and *dishwashing detergent* (sodium lauryl sulfate) *solutions* can be washed down the drain with plenty of water, provided your school drains are connected to a sanitary sewer system with a treatment plant.
- Set out a disposal container that is clearly labeled for the *rubbing alcohol solution*. Ensure that the pH is between 5 and 9, and then dilute it with 10 times as much water before pouring it down the drain.
- Set out a disposal container that is clearly labeled for the *milk of magnesia* (magnesium hydroxide) *solution*. Ensure that the pH is about 10 before pouring it down the drain.
- Set out a disposal container that is clearly labeled for the *chemical fertilizer solution*. Check federal, state, and local regulations regarding its proper disposal.

▶ Sample Data

Table 1 The Conductivity of Some Household Solutions

Beaker number	Household solution	Conductivity (μS/cm)	Ionic, covalent, or polyatomic?
1	Tap water	83	Covalent (with some ions present)
2	Table sugar	0	Covalent
3	Table salt	12 458	Ionic
4	Rubbing alcohol	0	Covalent
5	Milk of magnesia	1004	Ionic
6	Epsom salt	3684	Polyatomic (ionic and covalent)
7	Baking soda	8178	Polyatomic (ionic and covalent)
8	Dishwashing detergent	487	Polyatomic (ionic and covalent)
9	Chemical fertilizer	2197	Polyatomic (ionic and covalent)

Answers

▶ Analyzing Your Results

1. For the sample data provided:

Table 2 Classifying Household Solutions

No conductivity (0 μS/cm)	Solutions of table sugar and rubbing alcohol
Low conductivity (1–500 μS/cm)	Tap water and dishwashing detergent solution
High conductivity (>500 μS/cm)	Solutions of table salt, milk of magnesia, Epsom salt, chemical fertilizer, and baking soda

▶ Reaching Conclusions

2. The solutions with high conductivities listed in **Table 2** are most likely contributing to the conductivity because they contain ions that can conduct electricity. Tap water and dishwashing detergent might also be contributing slightly.
3. Tap water shows some conductivity because it contains some ions that can conduct electricity.
4. The table salt solution shows the highest conductivity.
5. Given that information, chemical fertilizer is probably the greatest contributor to the high conductivity of the river.
6. Most of the solutions of organic compounds tested do not conduct electricity. An exception is the dishwashing detergent solution. It can conduct electricity because it contains Na^+ and other ions.
7. See **Table 1** on the previous page.

▶ Defending Your Conclusions

8. Answers will vary. One possible recommendation would be to require the golf course to stop using chemical fertilizers and to use more natural products instead.

▶ Expanding Your Knowledge

1. Conductivity is an indicator of the total dissolved solids (TDS) present in a water sample and is often used by water quality experts as a general indicator of the natural or human inputs making their way into a body of water.
2. One way students might try to test the conductivity of soil is to add the soil to deionized water, shake the mixture, and then allow the mixture to stand for several hours. Then students can measure the conductivity of the water containing the dissolved minerals.

Investigating the Effect of Temperature on the Rate of a Reaction

MATERIALS (for each lab group)

- ✓ temperature probe
- ✓ CBL
- ✓ TI graphing calculator and black link cable
- ✓ water bath
- ✓ plastic test-tube rack
- ✓ 2 test tubes
- ✓ 10 mL graduated cylinders (2)
- ✓ warm tap water
- ✓ cold tap water
- ✓ 50 mL beakers (2)
- ✓ 1.0 M HCl
- ✓ 0.10 M $Na_2S_2O_3$
- ✓ stopwatch
- ✓ graph paper (optional)

Students should wear safety goggles, gloves, and laboratory aprons while performing this experiment.

Make sure you have Material Safety Data Sheets (MSDSs) available for all chemicals. Emphasize to students that hydrochloric acid is corrosive and can cause burns if it comes in contact with skin. If the solutions spill onto a student's gloves, the student should promptly remove the gloves and put on a fresh pair. If a chemical comes in contact with a student's bare skin, have the student rinse the affected area immediately and continuously with lukewarm water.

▶ Objectives

In this lab, students will

- ***perform and observe a reaction*** several times, at a different temperature each time.
- ***measure*** the time it takes for each reaction to be completed.
- ***make graphs*** of their data.
- ***determine*** how the rate of a chemical reaction varies with temperature.

▶ Planning for the Experiment

Organizing your class

Having two students in each lab group works best—one to note the time and/one to do the combining and observing.

Recommended time

1 lab period

Preparing the CBL System

- If possible, have a temperature probe, a CBL, and a graphing calculator connected for each group ahead of time. The calculators should be loaded with the Vernier PHYSCI program.

Preparing other materials

- You can purchase 1.0 M HCl or you can make it by adding 83 mL of concentrated HCl to a 1 L volumetric flask containing about 500 mL of deionized water and then filling to the mark with deionized water. **Wear safety goggles, gloves, and a laboratory apron. Be sure you add acid to water and not water to acid.**
- You can make 0.10 M $Na_2S_2O_3$ by dissolving 25 g of $Na_2S_2O_3$ pentahydrate in enough deionized water to make 1 L.
- Make sure the temperature of your warm tap water exceeds 40°C and that the temperature of your cold tap water does not exceed 10°C.

▶ Procedural Notes

General

- Students should transfer the solutions to test tubes and pour the contents of the test tubes back and forth over a paper-lined tray to catch spills.
- Have some sodium bicarbonate (baking soda) on hand to neutralize any spills.
- Tell students it is okay to run the reaction at a temperature close to the one asked for in the lab, as long as they record the exact temperature of the water in their data table.

Be aware that there is a possibility that some people may experience an allergic reaction to either sodium thiosulfate or sulfur. Allow any affected people to go outside to get some fresh air.

Caution students also not to breathe the acid's fumes.

Make sure students handle glassware with caution and dispose of any broken glass in a specially designated container.

Follow the disposal instructions given in the *Procedural Notes* section. Make sure that students wash their hands when they are done with the experiment.

- Make sure the reaction is not carried out at an extreme temperature. When heated to decomposition, sodium thiosulfate yields toxic fumes.
- You might want to have students run the reaction one time without timing it so that they can simply observe the formation of product. Then they will be better prepared when they see this happen during the first trial. Placing a dark material, such as black electrical tape, underneath the test tubes may help students see the sulfur forming.

Specifics

- **Step 3** Give students slightly more than 20 mL of each solution. Have students transfer the solutions to their work area on a paper-lined tray.
- **Step 15** Caution students to be very careful when rinsing the test tubes. Glass can be very slippery when it is wet.

DISPOSAL INSTRUCTIONS

- Set out a disposal container that is clearly labeled for the *reaction waste*. Filter, dry, and store the sulfur in a labeled reagent bottle for reuse. (If you will not reuse it, it must be buried in a landfill that is approved for chemical disposal.) Neutralize the filtrate with sodium carbonate (soda ash) or calcium hydroxide (lime) until the pH is between 6 and 8. You can then wash the solution down the drain with plenty of water, provided your school drains are connected to a sanitary sewer system with a treatment plant.
- Set out a disposal container that is clearly labeled for any leftover *hydrochloric acid solution*. Neutralize the solution with 1.0 M NaOH as required until the pH is between 6 and 8 before pouring it down the drain.
- Set out a disposal container that is clearly labeled for any leftover *sodium thiosulfate solution*. Allow the excess water to evaporate from the solution, filter the solids, and store them in a labeled reagent bottle for reuse. (It will be necessary to pulverize the crystals before reusing them.) The filtrate may be poured down the drain.

▶ Sample Data

Temperature and Reaction Time of Reactants

Trial	Temperature (°C)	Elapsed reaction time (s)
1	43	15
2	32	22
3	19	34
4	12	45

Answers

▶ Analyzing Your Results

1. Graphs will vary. For the sample data provided:

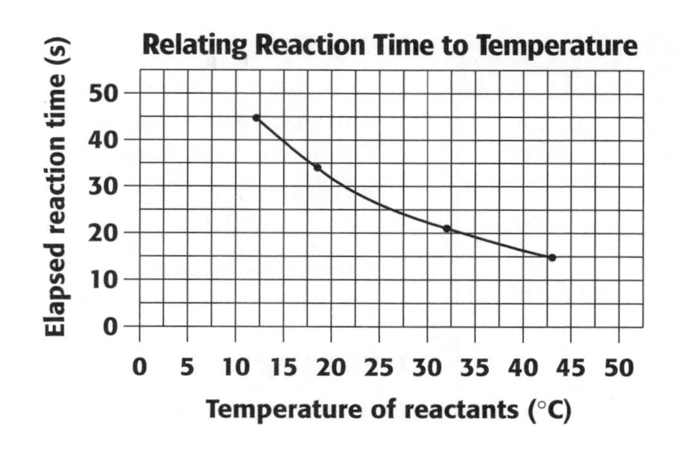

2. The higher the temperature of the reactants, the shorter the reaction time.
3. Answers will vary. Based on the graph of the sample data, the reaction would take about 55–60 s to occur at 5°C. The reaction would take about 10–12 s to occur at 50°C.
4. No; the water in the solutions would be frozen, so the reactants would not be able to chemically combine.

▶ Reaching Conclusions

5. Answers will vary. Based on the graph of the sample data, the reaction would have to be run at about 35°C or higher to be complete in 20 s or less.
6. Cooling a chemical in the refrigerator slows the rate of decomposition and therefore extends its shelf life.

▶ Defending Your Conclusions

7. Answers will vary. One possible answer is that the student could perform the experiment again, testing the actual reaction used to develop film (reacting silver bromide with sodium thiosulfate).

▶ Expanding Your Knowledge

1. Make sure you supervise students as they perform the experiment again. Students should find that the more dilute the hydrochloric acid solution is, the longer it takes for the reaction to occur.

Determining the Concentration of an Ionic Solution

MATERIALS (for each lab group)

- ✓ conductivity probe (and adapter cable)
- ✓ CBL
- ✓ TI graphing calculator and black link cable
- ✓ 100 mL beakers (8)
- ✓ several NaCl solutions of different concentrations
- ✓ 100 mL graduated cylinder
- ✓ deionized water
- ✓ rinse bottle filled with deionized water
- ✓ large beaker
- ✓ tissue paper
- ✓ ring stand
- ✓ small clamp
- ✓ graph paper (optional)

Students should wear safety goggles, gloves, and laboratory aprons while performing this experiment.

If a chemical comes in contact with a student's bare skin, have the student rinse the affected area immediately and continuously with lukewarm water.

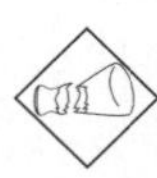

Make sure students handle glassware with caution and dispose of any broken glass in a specially designated container.

▶ Objectives

In this lab, students will

- ***measure*** the conductivities of several NaCl solutions of known concentrations.
- ***relate*** their conductivity measurements to the concentrations of the solutions.
- ***determine*** whether two solutions of NaCl have been labeled with the right concentrations.

▶ Planning for the Experiment

Organizing your class

Having four students in each lab group works best.

Recommended time

1 lab period

Preparing the CBL System

- If possible, have a conductivity probe, a CBL, and a graphing calculator connected for each group ahead of time. The calculators should be loaded with the Vernier PHYSCI program.

Preparing other materials

- Make a NaCl stock solution by dissolving 20.0 g of NaCl in enough deionized water to make 1 L.
- Use the following information to prepare the standards:

Standard concentration (mass percent)	Preparation instructions
0.50% NaCl	Add 250 mL of the stock solution to 750 mL of deionized water.
0.40% NaCl	Add 200 mL of the stock solution to 800 mL of deionized water.
0.30% NaCl	Add 150 mL of the stock solution to 850 mL of deionized water.
0.20% NaCl	Add 100 mL of the stock solution to 900 mL of deionized water.
0.10% NaCl	Add 50 mL of the stock solution to 950 mL of deionized water.

- In this lab, students will discover that the two saline solutions have been mixed up and are not labeled correctly. Make one of the unknown solutions by dissolving 2.5 g of NaCl in enough deionized water to make 1 L. Label this solution incorrectly as 0.45 percent NaCl. (It is really 0.25 percent NaCl.)

The sodium chloride solutions can be washed down the drain with plenty of water, provided your school drains are connected to a sanitary sewer system with a treatment plant. Make sure that students wash their hands when they are done with the experiment.

- Make the other unknown solution by dissolving 4.5 g of NaCl in enough deionized water to make 1 L. Label this solution incorrectly as 0.25 percent NaCl. (It is really 0.45 percent NaCl.)

▸ Procedural Notes

General

- If you have time constraints, students can test as few as three standard solutions.
- Emphasize that it is very important for students to label their samples clearly.
- Have students transfer and keep all of their solutions on a paper-lined tray to catch spills.
- Make sure students use care when handling the conductivity probe. When testing a sample, the probe should not hit the sides or bottom of the beaker.
- Remind students to rinse the probe with deionized water between samples to avoid contamination.

Specifics

- **Step 10** Students should hold the probe the entire time without letting it rest on the sides of the beaker. It might be a good idea to mount each probe on a ring stand with a small clamp. Then students can lift and lower the clamp as needed to test their samples.

▸ Sample Data

Table 1 Conductivity Data of Standards

Beaker number	Concentration of standard solution	Conductivity (μS/cm)
1	0.50% NaCl	10 000
2	0.40% NaCl	8000
3	0.30% NaCl	6000
4	0.20% NaCl	4000
5	0.10% NaCl	2000
6	0% NaCl (deionized water)	0

Table 2 Conductivity Data of Unknowns

Beaker number	Concentration of solution	Conductivity (μS/cm)
7	Unknown (Labeled as "0.25% NaCl.")	9000
8	Unknown (Labeled as "0.45% NaCl.")	5000

Answers

▶ Analyzing Your Results

1. Graphs will vary. For the sample data provided:

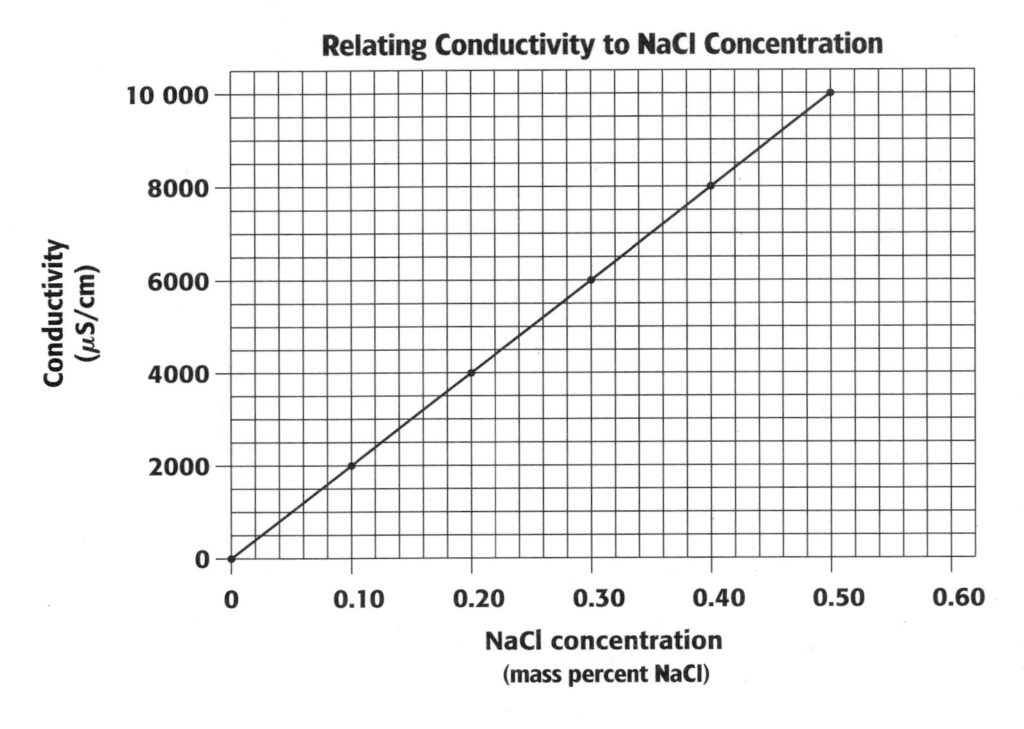

2. The solution in beaker 7 has a concentration of about 0.45 percent NaCl, while the solution in beaker 8 has a concentration of about 0.25 percent NaCl.

▶ Reaching Conclusions

3. No, the two solutions were switched.
4. Although it is possible that the graph may continue to be linear for higher concentrations, that assumption would have to be verified by testing the conductivities of more concentrated standards.

▶ Defending Your Conclusions

5. This experiment will not work because sugar is made of molecules that do not conduct electricity.

▶ Expanding Your Knowledge

1. One hypothesis might be "Rainfall will increase the conductivity of the water." Supervise students as they test their hypotheses over a period of several weeks. Remind students to be careful when collecting their water samples so that they don't slip and hurt themselves.
2. Students should verify that the higher the conductivity reading of the soup, the greater the mass of sodium listed on the soup's nutritional information label.

Determining the Effective Half-life of Iodine-131 in the Human Body

MATERIALS (for each student)

✓ TI graphing calculator (or graphing software or graph paper and a pencil)

Students should wear safety goggles while performing this experiment.

If students are working in the laboratory, they should also wear gloves and laboratory aprons.

Make sure that students wash their hands when they are done with the experiment.

▶ Objectives

In this lab, students will

- ***make a graph*** from research data that shows the decay activity of the radioactive isotope iodine-131 in the human body.
- ***analyze*** the graph to find out how iodine-131 is taken up by the thyroid gland.
- ***calculate*** the effective half-life of iodine-131 in the body.

▶ Planning for the Experiment

Organizing your class

Have students either work alone or in small groups.

Recommended time

1 lab period

Preparing materials

- This lab's procedure is written for a TI graphing calculator, but any graphing calculator or graphing software will work. Students can also use graph paper.

▶ Procedural Notes

Specifics

- **Step 4** Students must name the lists, or columns, they will use to enter data.
- **Step 8** Students must label the Xlist and Ylist with the exact same names that were used in step 4.

▶ Sample Data

Decay Activity for Iodine-131 in the Human Body

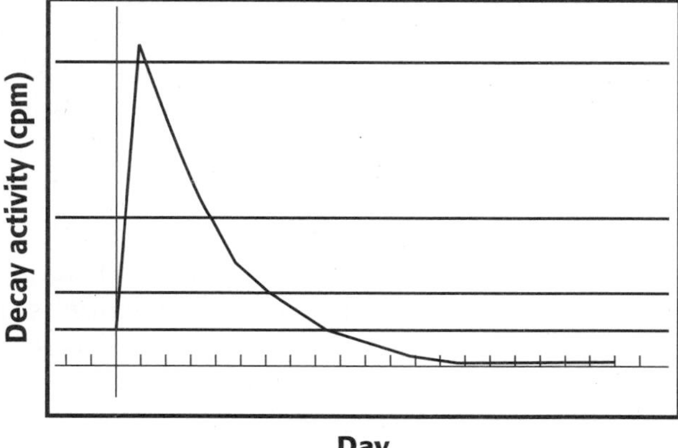

Answers

▶ Analyzing Your Results

1. The concentration of iodine-131 is very low at the start of the experiment and increases to a maximum on day 1. After day 1, the concentration of iodine-131 steadily declines.

2. The maximum concentration occurred sometime during the first 24 hours.

3. The following estimates can be obtained from the graph of the data.

Calculating the Effective Half-life of Iodine-131

Decay rate (cpm)	Time (days)	Elapsed time to halve the decay rate (days)
10 000	1.3	
5000	3.9	2.6
2500	6.5	2.6
1250	8.9	2.4
Average half-life		2.5

4. Answers may vary. For the sample data provided:

$$\text{time} = 3.9 \text{ days} - 1.3 \text{ days} = 2.6 \text{ days}$$

5. Answers may vary. For the sample data provided:

$$\text{time} = 6.5 \text{ days} - 3.9 \text{ days} = 2.6 \text{ days}$$
$$\text{time} = 8.9 \text{ days} - 6.5 \text{ days} = 2.4 \text{ days}$$

6. Answers may vary. For the sample data provided:

$$\text{average half-life} = \frac{2.6 \text{ days} + 2.6 \text{ days} + 2.4 \text{ days}}{3} = 2.5 \text{ days}$$

▶ Reaching Conclusions

7. The concentration of iodine-131 was initially low because the drug was given orally and it took some time to be assimilated into the patient's digestive track and to circulate through the bloodstream. Once the iodine-131 was taken up by the patient's thyroid gland, the concentration increased quickly. The concentration of iodine-131 slowly decreased over time as the patient's body eliminated it.

8. For this experiment, the effective half-life of iodine-131 in the thyroid gland was determined to be approximately 2.5 days. This value is much lower than the actual value because iodine-131 in the human body not only decays normally but is also eliminated.

9. The graph would have looked similar, except that the decay rates would have been lower.

▶ Defending Your Conclusions

10. Answers will vary but might include that if the data had shown a great deal of variation, doctors might have to run a test on an individual before prescribing a specific dose instead of using the averaged data to estimate an effective dose.

▶ Expanding Your Knowledge

1. There is a great deal of information that describes the elevated levels of radioactive iodine that were present as a result of the accident at Chernobyl in 1986. Elevated levels of iodine-131 were detected throughout the Northern Hemisphere after the accident. Infants (via their mother's milk) and children received the highest exposures.
2. Divide the class into groups. Each group can research one of the proposed ways to store radioactive waste. Students should evaluate the advantages and disadvantages of the method researched. As each group presents its findings, other members of the class should provide counterexamples. The presenting group should be able to respond to the arguments made by their classmates and defend its position.

Additional Notes

LAB
8 TEACHER'S NOTES FORCES

Determining Your Acceleration on a Bicycle

MATERIALS (for each lab group)

- ✓ force sensor (and adapter cable)
- ✓ CBL
- ✓ TI graphing calculator and black link cable
- ✓ multi-gear bicycle
- ✓ table support
- ✓ object with a mass of 2–6 kg
- ✓ string
- ✓ sturdy table
- ✓ short piece of rope
- ✓ bathroom scale
- ✓ graph paper (optional)
- ✓ calculator (optional)

Students should wear safety goggles and gloves while performing this experiment.

If students are working in the laboratory, they should also wear laboratory aprons.

Do not let students ride the bicycles inside the school building or inside the lab.

Emphasize to students that it is very important that they keep their fingers, hair, clothing, and jewelry away from the spokes and gear-wheels.

NOTE: Students may not be able to get a 0 reading on the force sensor. They should adjust the sensor to get as close to 0 as possible.

▶ Objectives

In this lab, students will

- ***design*** a system that can simulate the forces acting on a bicycle as it is being ridden.
- ***measure*** the forward, or impelling, force acting on the bicycle for each gear.
- ***calculate*** the acceleration that is possible for each gear by using Newton's second law of motion.
- ***determine*** which gears provide the greatest acceleration.

▶ Planning for the Experiment

Organizing your class

Having at least three students in each lab group works best.

Recommended time

1–2 lab periods

Preparing the CBL System

- If possible, have a force sensor, a CBL, and a graphing calculator connected for each group ahead of time. The calculators should be loaded with the Vernier PHYSCI program.
- This lab's procedure is written for a student force sensor, which can measure forces up to 10 N. If dual-range force sensors are used, the CBL System will need to be set up differently, as described in the *Procedural Notes* section.

▶ Procedural Notes

General

- When students are collecting data, they should not push or pull on the bicycle, as this will affect their results.

Specifics

- **Steps 5 and 6** If students are using dual-range force sensors, they should substitute the following steps for steps 5 and 6:
  5. Go to the MAIN MENU, and select SET UP PROBES. Enter "1" as the number of probes. Select FORCE from the SELECT PROBE menu. Enter "1" as the channel number.
  6. Select USE STORED from the CALIBRATION menu. Select DUAL-RANGE from the FORCE PROBE menu, then select DUAL-RANGE 10 N (or 50 N, depending on the masses you provide). Set the switch on the sensor to 10 N (or 50 N).

Make sure that students wash their hands when they are done.

▶ Sample Data

Table 1 Forces and Masses of Objects

Object	Weight (lb)	Force (N)	Mass (kg)	Ratio of student's force to hanging object's force
Bicycle and student	153	680	69	
Student	102	453		
Hanging object		19.6	2.00	23.1

Table 2 Force and Acceleration Measurements and Calculations

Front gear-wheel	Rear gear-wheel	Impelling force hanging object produces (N)	Student's impelling force (N)	Student's acceleration on the bicycle (m/s²)
Smallest	S1 (low)	9.00	208	3.0
	S2	8.20	189	2.7
	S3	7.50	173	2.5
	S4	6.30	146	2.1
	S5	5.50	127	1.8
Middle	M1 (low)	6.00	139	2.0
	M2	5.30	122	1.8
	M3	4.70	109	1.6
	M4	4.10	94.7	1.4
	M5	3.50	80.9	1.2
Largest	L1 (low)	4.20	97.0	1.4
	L2	3.80	87.8	1.3
	L3	3.40	78.5	1.1
	L4	2.90	67.0	0.97
	L5	2.50	57.8	0.84

Answers

▶ Analyzing Your Results

1. Answers will vary. For the sample data provided:

$$\text{student's force} = 102\ \cancel{\text{lb}} \times \frac{1\ \text{N}}{0.225\ \cancel{\text{lb}}} = 453\ \text{N}$$

$$\text{combined force of student and bicycle} = 153\ \cancel{\text{lb}} \times \frac{1\ \text{N}}{0.225\ \cancel{\text{lb}}} = 680\ \text{N}$$

2. Answers will vary. For the sample data provided:

$$\text{combined mass of student and bicycle} = \frac{680\ \text{N}}{9.8\ \text{m/s}^2} = 69\ \text{kg}$$

LAB 8 NOTES

3. Answers will vary. For the sample data provided:

$$\textbf{hanging object's force} = 2.00 \text{ kg} \times 9.8 \text{ m/s}^2 = 19.6 \text{ N}$$

4. Answers will vary. For the sample data provided:

$$\textbf{ratio of forces} = \frac{453 \text{ N}}{19.6 \text{ N}} = 23.1$$

5. Answers will vary. The calculation for the first gear (S1) is shown below. Answers for all the gears are listed in **Table 2.**

$$\textbf{student's impelling force in gear S1} = 9.00 \text{ N} \times 23.1 = 208 \text{ N}$$

6. Answers will vary. The calculation for the acceleration of a 102 lb student in first gear (S1) is shown below. Answers for all the gears are listed in **Table 2.**

$$\textbf{student's acceleration in gear S1} = \frac{208 \text{ N}}{69 \text{ kg}} = 3.0 \text{ m/s}^2$$

7. Graphs will vary. For the sample data provided:

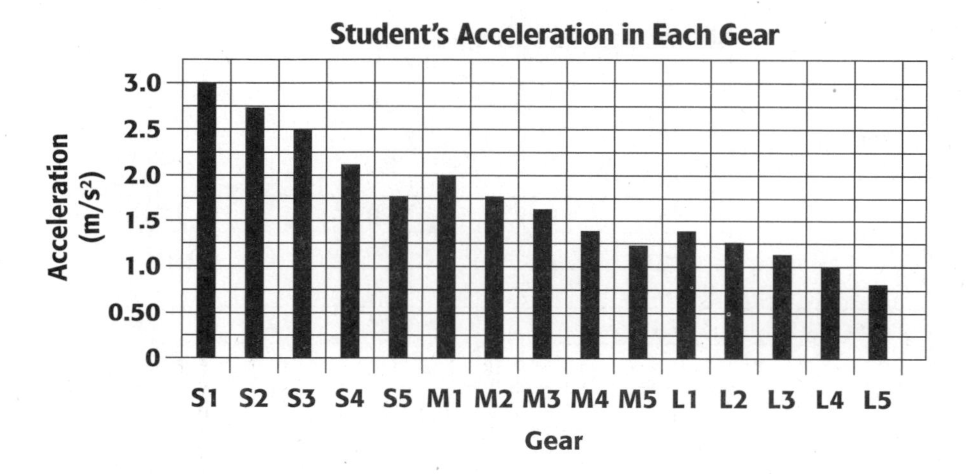

▶ Reaching Conclusions

8. Answers will vary. For the sample data provided:

S1, S2, S3, S4, M1, S5, M2, M3, M4, L1, L2, M5, L3, L4, L5

9. Answers will vary. Based on the sample data provided, gear S5 overlaps with gear M1, and gear M5 overlaps with gear L2.

▶ Defending Your Conclusions

10. The actual force exerted by a rider would be less than the rider's weight, so the rider's acceleration would also be less.

▶ Expanding Your Knowledge

1. Answers will vary for each bicycle and each gear.

Determining Which Ramp Is More Efficient

MATERIALS (for each lab group)

- ✓ force sensor (and adapter cable)
- ✓ CBL
- ✓ TI graphing calculator and black link cable
- ✓ sturdy chair
- ✓ board approximately three times as long as the chair's height
- ✓ several heavy books
- ✓ rolling cart
- ✓ meterstick
- ✓ string

Students should wear safety goggles and gloves while performing this experiment.

If students are working in the laboratory, they should also wear laboratory aprons.

Make sure that students place the rolling carts on the lab tables when they are not being used. Don't let students leave the carts on the floor where someone might step on them.

Make sure that students wash their hands when they are done with the experiment.

NOTE: Students may not be able to get a 0 reading on the force sensor. They should adjust the sensor to get as close to 0 as possible.

▶ Objectives

In this lab, students will

- ***measure*** the force required to pull an object up a ramp under different conditions.
- ***calculate*** work input and work output.
- ***compare*** the mechanical advantage and efficiency of two ramps that have different lengths.

▶ Planning for the Experiment

Organizing your class

Having four students in each lab group works best.

Recommended time

1 lab period

Preparing the CBL System

- If possible, have a force sensor, a CBL, and a graphing calculator connected for each group ahead of time. The calculators should be loaded with the Vernier PHYSCI program.
- This lab's procedure is written for a student force sensor, which can measure forces up to 10 N. If dual-range force sensors are used, the CBL System will need to be set up differently, as described in the *Procedural Notes* section.

Preparing other materials

- Make sure that the boards are fairly smooth.
- Provide carts that have a flat surface on the top and wheels on the bottom. If the carts are heavier than 10 N, have students use dual-range force sensors instead of student force sensors.

▶ Procedural Notes

General

- Be sure to keep a direct path to the exit clear of any obstructions in case of an emergency.

Specifics

- **Steps 4 and 5** If students are using dual-range force sensors, they should substitute the following steps for steps 4 and 5:
  4. Go to the MAIN MENU, and select SET UP PROBES. Enter "1" as the number of probes. Select FORCE from the SELECT PROBE menu. Enter "1" as the channel number.
  5. Select USE STORED from the CALIBRATION menu. Select DUAL-RANGE from the FORCE PROBE menu. Then select DUAL-RANGE 10 N (or 50 N, depending on the rolling carts you provide). Set the switch on the sensor to 10 N (or 50 N).

- **Step 8** If the weight of the cart is too heavy for the force sensor to measure, students can determine the cart's mass in kilograms on a balance, and then multiply by 9.8 m/s^2 to find the cart's weight in newtons.
- **Step 11** Emphasize to students that it is very important that they use a consistent amount of force as they pull the cart up the ramp.

▶ Sample Data

Table 1 Work Output for Both Ramps

Height of chair (to seat) (m)	Weight of cart (N)	Work output (J)
0.45	5.3	2.4

Table 2 Work Input, Mechanical Advantage, and Efficiency of Each Ramp

Ramp length (m)	Sliding or rolling the cart?	Force (N)	Work input (J)	Mechanical advantage	Efficiency (%)
1.40	Sliding	3.0	4.2	3.1	57%
	Rolling	2.0	2.8		86%
0.90	Sliding	4.0	3.6	2.0	67%
	Rolling	3.0	2.7		89%

Answers

▶ Analyzing Your Results

1. Answers will vary. For the sample data provided:

$$\textbf{work output} = 0.45 \text{ m} \times 5.3 \text{ N} = 2.4 \text{ J}$$

2. Answers will vary. The calculation for sliding the cart up the longer ramp is shown below. Answers for both ramps are listed in **Table 2.**

$$\textbf{work input (longer ramp/sliding the cart)} = 3.0 \text{ N} \times 1.40 \text{ m} = 4.2 \text{ J}$$

3. Answers will vary. The calculation for the longer ramp is shown below. Answers for both ramps are listed in **Table 2.**

$$\textbf{mechanical advantage (longer ramp)} = \frac{1.40 \cancel{\text{m}}}{0.45 \cancel{\text{m}}} = 3.1$$

4. Answers will vary. The calculation for sliding the cart up the longer ramp is shown below. Answers for both ramps are listed in **Table 2.**

$$\textbf{percent efficiency (longer ramp/sliding the cart)} = \frac{2.4 \cancel{\text{J}}}{4.2 \cancel{\text{J}}} \times 100 = 57\%$$

▶ Reaching Conclusions

5. The longer ramp has a greater mechanical advantage. The longer a ramp is, the greater its mechanical advantage. Speedy Shipping Company needs a ramp with a greater mechanical advantage.

6. The shorter ramp is more efficient both when sliding and when rolling the cart because the cart doesn't have to be moved as far as it does when using the longer ramp.

7. In each case, the difference between work input and work output was used to overcome friction.

8. If the truck driver has a good rolling cart, the friction will be negligible. The length of ramp needed can be calculated by the following equation.

$$\textbf{length of ramp} = \frac{\textbf{1010 }\cancel{\textbf{N}} \times \textbf{1.0 m}}{\textbf{250 }\cancel{\textbf{N}}} = \textbf{4.0 m}$$

The ramp needs to be at least 4.0 m long.

▶ Defending Your Conclusions

9. The results are still valid because pulling a light object still demonstrates the relationship that exists between ramp length, mechanical advantage, and efficiency.

▶ Expanding Your Knowledge

1. Students can visit local retailers or research product information on the Internet. Some states even post information about specific brand name appliances.

Additional Notes

LAB 9 NOTES

LAB

10 TEACHER'S NOTES **HEAT AND TEMPERATURE**

Determining the Better Insulator for Your Feet

MATERIALS (for each lab group)

- ✓ 2 temperature probes
- ✓ CBL
- ✓ TI graphing calculator and black link cable
- ✓ 2 plastic beverage bottles
- ✓ 2 rubber stoppers that fit temperature probes and bottle openings
- ✓ cotton sock
- ✓ wool sock
- ✓ warm tap water at 40–45°C (you may need a hot plate and a metal pan)
- ✓ balance
- ✓ cold tap water
- ✓ paper towels
- ✓ stopwatch

Students should wear safety goggles and gloves while performing this experiment.

If students are working in the laboratory, they should also wear laboratory aprons.

Make sure that students wash their hands when they are done with the experiment.

▶ Objectives

In this lab, students will

- ***model*** a hiker's feet in wet conditions using bottles of water covered with wet socks.
- ***measure*** the temperature changes.
- ***calculate*** the energy transferred away from each "foot" as heat in joules by using the specific heat equation.
- ***compare*** the effectiveness of the two socks as insulators in wet conditions.
- ***evaluate*** which material insulates better.

▶ Planning for the Experiment

Organizing your class

Having two to four students in each lab group works best.

Recommended time

1 lab period

Preparing the CBL System

- If possible, have the temperature probes, a CBL, and a graphing calculator connected for each group ahead of time. The calculators should be loaded with the Vernier PHYSCI program.

Preparing other materials

- Remove any labels or other insulating materials from the bottles before using them. The bottles must be small enough so that socks can fit over them snugly.
- Provide identical cotton socks and identical wool socks for each group.
- Make sure the rubber stoppers you use fit the temperature probes and the openings of the bottles. To make the temperature probes easier to insert into the rubber stoppers, lubricate the openings with grease.

▶ Procedural Notes

General

- Make sure the socks do not cause the bottles to become unstable if the bottles are free-standing.
- If you have an extra temperature probe, stopper, and bottle, you might want to prepare and collect data for a control "foot" not covered by a wet sock. That way, students can compare the temperature changes that occur for a bare "foot" with those that occur for the "feet" covered with socks.

Specifics

- **Step 8** To save time, make marks on the bottles indicating the appropriate level of water. Another option would be to have students experiment to find the appropriate level the day before the lab.
- **Step 9** Emphasize to students that the outsides of the bottles must be dry when they are measuring the mass of the bottles filled with water.

▶ Sample Data

Table 1 Mass of Warm Water Used to Make "Feet"

Bottle number	Mass of bottle (g)	Mass of bottle and water (g)	Mass of water (g)	Mass of water (kg)
1	4.2	467.9	463.7	0.4637
2	4.4	468.4	464.0	0.4640

Table 2 Temperature Changes of Each "Foot"

Bottle number/sock material	Initial temp. (°C)	After 4 min (°C)	After 8 min (°C)	After 12 min (°C)	Final temp. (°C)	Δt (°C)	Energy transferred as heat (J)
1/cotton	45.9	43.5	38.9	36.5	34.7	−11.2	-2.17×10^4
2/wool	42.8	41.0	39.6	37.5	38.5	−4.3	-8.35×10^3

Answers

▶ Analyzing Your Results

1. Answers will vary. The calculation for bottle 1 is shown below. Answers for both bottles are listed in **Table 1.**

$$\textbf{mass of water in bottle 1} = 467.9\text{ g} - 4.2\text{ g} = 463.7\text{ g}$$

2. Answers will vary. The calculation for bottle 1 is shown below. Answers for both bottles are listed in **Table 1.**

$$\textbf{mass of water in bottle 1 (kg)} = 463.7\text{ g} \times \frac{1\text{ kg}}{1000\text{ g}} = 0.4637\text{ kg}$$

3. Answers will vary. The calculation for bottle 1 is shown below. Answers for both bottles are listed in **Table 2.**

$$\Delta t \text{ for bottle 1} = 34.7°\text{C} - 45.9°\text{C} = -11.2°\text{C}$$

Remind students that answers are negative because the temperature dropped over time.

4. Answers will vary. The calculation for bottle 1 is shown below. Answers for both bottles are listed in **Table 2.** (**Note:** This problem is simplified by giving the value of the specific heat of water as 4186 J/kg•K—the value at 25°C. More accurate specific heat values of water in the 30–45°C temperature range can be found in the CRC or another reference book.)

$$\textbf{energy transferred} = \frac{4186\ \text{J}}{\text{kg} \cdot \text{K}} \times 0.4637\ \text{kg} \times (-11.2°\text{C}) = -2.17 \times 10^4\ \text{J}$$

Remind students that answers are negative because energy is released and that degrees Celsius and Kelvin cancel because a temperature change of 1°C is equal to that of 1 K.

▶ Reaching Conclusions

5. Cotton allowed more energy to be transferred away from the "foot" as heat.
6. Answers will vary. About 2.6 times as much energy was transferred away from the "foot" wearing the cotton sock as was transferred away from the "foot" wearing the wool sock.
7. Hikers should wear wool socks when hiking in the cold rain because less energy will be transferred away from their feet as heat, causing them to stay warmer.

▶ Defending Your Conclusions

8. For a real foot, in which energy in the form of heat is constantly being replaced by circulating blood, the amount of energy transferred away as heat would be much greater than that calculated in this experiment. However, the insulating abilities of the socks can still be compared using this model.

▶ Expanding Your Knowledge

1. A classroom contest can be conducted in which teams of students design a layer of insulation around a bottle using 1 kg of any material they choose.
2. After students do their research, have them give a presentation on their findings.
3. Students can test different varieties of socks or other garments made from synthetic materials to compare them with wool and cotton.
4. The structure of wool provides maximum insulation. Students can go to the library or use the Internet to learn more about the structure of wool.

Tuning a Musical Instrument

MATERIALS (for each lab group)

- ✓ microphone (CBL-compatible)
- ✓ CBL
- ✓ TI graphing calculator and black link cable
- ✓ TI-Graph Link cable (optional)
- ✓ computer with a printer (optional)
- ✓ calculator (optional)
- ✓ tuning forks for comparison (optional)
- ✓ any stringed musical instrument (guitar, mandolin, banjo, violin, etc.)

Students should wear safety goggles while performing this experiment.

If students are working in the laboratory, they should also wear gloves and laboratory aprons.

Remind students to treat the musical instruments with care, especially if they are borrowing someone else's instrument. Also remind them not to use excessive force when plucking the strings.

Make sure that students wash their hands when they are done with the experiment.

▸ Objectives

In this lab, students will

- ***relate*** the pitch of a musical note to its frequency.
- ***apply*** the concept of wave interference to tune an instrument using an electronic device.
- ***compare*** their ability to tune an instrument using an electronic device with their ability to tune an instrument by ear.

▸ Planning for the Experiment

Organizing your class

Having two to four students in each lab group works best. Ideally, each group should have at least one member that can play a stringed musical instrument. It is best to do the experiment outside or in a very large (and quiet) room where students can spread out so that each group has relatively quiet surroundings.

Recommended time

1 lab period

Preparing the CBL System

- If possible, have a CBL-compatible microphone, a CBL, and a graphing calculator connected for each group ahead of time. The calculators should be loaded with the Vernier PHYSCI program.
- If a note is played too loudly, a ragged waveform may result. The day before the lab, you might want to have students experiment with the microphone placement and volume to determine how to produce the smoothest waveform possible.

Preparing other materials

- If you do not have a TI-Graph Link cable or a computer with a printer, students can sketch the graphs in their lab reports instead of printing them out.
- Find out who in your class can play a stringed musical instrument. Organize who will be bringing in their instruments several days before the lab. You only need to have one stringed instrument per lab group.
- If you do not have access to stringed musical instruments, you can use tuning forks or electronic keyboards instead. Although these cannot be tuned like a stringed instrument, they can be used to observe beat frequencies. For best results, have students use tuning forks with frequencies that differ by approximately 30 Hz, and enter "0.003" as the time between samples in seconds. If students are using electronic keyboards, have them choose notes such as the middle C and E.

▶ Procedural Notes

General

- It is important that students hold the microphone close to the instrument before playing the note or notes and then quickly press ENTER on the calculator. The sample time is very small. If students press ENTER and then try to move the microphone to the sound, they will miss the sound.
- Encourage students to choose pitches with frequencies of less than 400 Hz. This would correspond to notes on the lowest three strings of a guitar.
- When students are graphing two sounds at once, it is important that both sounds be played with equal volume. If students are using stringed instruments, remind them to hold the microphone in front of the instrument and pluck both strings equally hard. If students are using tuning forks, they should hold both forks the same distance from the microphone.

Specifics

- **Step 7** Students should count the number of complete waves there are from the first peak to the last peak.
- **Step 14** The time between samples may have to be adjusted, depending on the difference in frequency of the two sounds being played. The 0.01 s sampling interval should work well for pitches that are within 3 or 4 Hz of each other. To observe beats for two sounds that are much farther apart, the sampling interval should be reduced. If students need to modify the sampling interval, they should select the MODIFY SETUP command in the USE TIME SETUP menu.
- **Step 17** Students should count the number of complete cycles (from one maximum to the next) there are from the first maximum to the last maximum.

▶ Sample Data

Table 1 Two Different Sounds Played Separately (Sounds 1 and 2)

Sound	Amplitude (V)	Time of first peak (s)	Time of last peak (s)	Elapsed time (s)	Number of waves between first and last peaks	Period (s)	Frequency (Hz)
1	0.55	0.00250	0.01778	0.01528	4	0.00382	262
2	0.61	0.00290	0.01990	0.01700	5	0.00340	294

Table 2 Two Sounds Out of Tune that Are Played Together (Sound 3)

Maximum amplitude (V)	Time of first maximum (s)	Time of last maximum (s)	Elapsed time (s)	Number of cycles between first and last maximums	Beat period (s)	Frequency (Hz)
0.72	0.01950	0.09010	0.07060	2	0.03530	28.33

Answers

▶ Analyzing Your Results

1. Answers will vary. The calculation for sound 1 is shown below. Answers for sounds 1 and 2 are listed in **Table 1.**

$$\textbf{elapsed time for sound 1} = 0.01778 \text{ s} - 0.00250 \text{ s} = 0.01528 \text{ s}$$

2. Answers will vary. For the sample data provided:

$$\textbf{elapsed time for sound 3} = 0.09010 \text{ s} - 0.01950 \text{ s} = 0.07060 \text{ s}$$

3. Answers will vary. The calculation for sound 1 is shown below. Answers for sounds 1 and 2 are listed in **Table 1.**

$$\textbf{sound 1 period} = \frac{0.01528 \text{ s}}{4} = 0.00382 \text{ s}$$

4. Answers will vary. For the sample data provided:

$$\textbf{sound 3 beat period} = \frac{0.07060 \text{ s}}{2} = 0.03530 \text{ s}$$

5. Answers will vary. The calculation for sound 1 is shown below. Answers for all three sounds are listed in **Tables 1** and **2.**

$$\textbf{frequency of sound 1} = \frac{1}{0.00382 \text{ s}} = 262 \text{ Hz}$$

▶ Reaching Conclusions

6. The sound with the higher pitch has the higher frequency.
7. The maximums and peaks on the graphs show times when constructive interference occurred. The amplitude is greater because it is the sum of the amplitudes of the two individual sound waves.
8. The places on the graph where the amplitude nears zero show times when destructive interference occurred. The amplitude is smaller because the crest of one wave is added to the trough of the other wave.
9. Answers will vary. The quality of the microphone might affect the students' answers.

▶ Defending Your Conclusions

10. No, the electronic devices musicians use just simplify the process and require less analysis.

▶ Expanding Your Knowledge

1. Arrange for a piano tuner to visit your classroom to demonstrate the tuning process to your students.

Choosing a Pair of Sunglasses

MATERIALS (for each lab group)
- ✓ light sensor
- ✓ CBL
- ✓ TI graphing calculator and black link cable
- ✓ desk lamp
- ✓ piece of glass or aluminum foil
- ✓ ring stand
- ✓ clamp
- ✓ several different pairs of sunglasses, some having polarized lenses
- ✓ graph paper (optional)

Students should wear safety goggles and gloves while performing this experiment.

If students are working in the laboratory, they should also wear laboratory aprons.

Make sure students do not touch the light source or look directly at it. Remind students to allow ample time for the lamp to cool before they handle it.

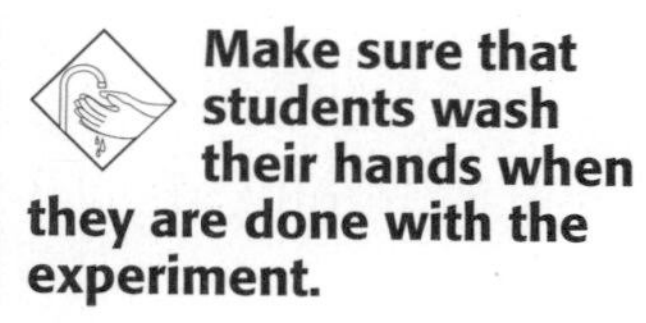

Make sure that students wash their hands when they are done with the experiment.

▶ Objectives

In this lab, students will
- ***measure*** the intensity of light that is reflected from a smooth, flat surface.
- ***compare*** several pairs of sunglasses for their ability to block reflected light while still allowing someone to see well.
- ***evaluate*** their data to determine which pair of sunglasses they would recommend that people wear.

▶ Planning for the Experiment

Organizing your class

Having two to four students in each lab group works best.

Recommended time

1 lab period

Preparing the CBL System

- If possible, have a light sensor, a CBL, and a graphing calculator connected for each group ahead of time. The calculators should be loaded with the Vernier PHYSCI program.

Preparing other materials

- Arrange for students to bring in sunglasses on the day of the lab. It might be a good idea to have darkened glass or colored cellophane on hand for those that do not have sunglasses or forget to bring them. Have a piece of polarized glass on hand for each group in case students do not have sunglasses with polarized lenses.
- The desk lamp should be equipped with a 60 W light bulb.
- Give each group a hot pad or a padded glove in case they need to adjust the lamp while it is hot.
- If a piece of glass or aluminum foil is not available, a flat tray of water will also work.

▶ Procedural Notes

General

- Remind students to check the intensity of the reflected light before they test each pair of sunglasses.

Specifics

- **Steps 9 and 10** Be sure that students position the sunglasses so that the light passes through one of the lenses and not between the lenses through the nose piece.
- **Step 12** Do not let students look directly at the light bulb or at the sun regardless of whether they are wearing sunglasses or not.

▶ Sample Data

Sunglass Data

Light intensity reading without sunglasses	Brand/ style of sunglasses	Light intensity reading with sun-glasses horizontal	Light intensity reading with sun-glasses vertical	Maximum difference in light intensity readings
0.258	Brand X	0.165	0.163	0.095
0.248	Brand Y	0.056	0.058	0.192
0.257	Brand Z	0.078	0.112	0.179

Notes in lab report (step 11):
Brand X: Nonpolarized; yellow, reflective coating
Brand Y: Nonpolarized, brown
Brand Z: Polarized, gray

Answers

▶ Analyzing Your Results

1. Answers will vary. The calculation for Brand X is shown below. Answers for all three pairs of sunglasses are listed in the table above.

maximum difference in readings for Brand X = 0.258 − 0.163 = 0.095

2. Graphs will vary. For the sample data provided:

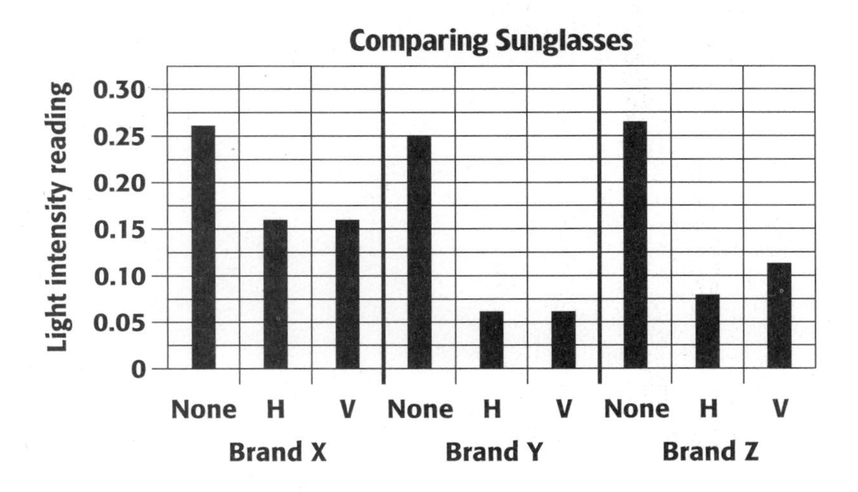

▶ Reaching Conclusions

3. Answers will vary. For the sample data provided, only the Brand Z sunglasses were polarized. Students should be able to read fine print while wearing polarized sunglasses because the lenses are not dark enough to hinder reading.

4. Answers will vary. For the sample data provided, Brand Y sunglasses were able to reduce the light's intensity the most, followed by Brand Z and then Brand X.
5. Answers will vary. Based on the sample data, Brand Y sunglasses blocked the most light, but they did not allow for reading fine print as well as Brand Z. Therefore, Brand Z is the best pair of sunglasses overall.

▶ Defending Your Conclusions

6. Answers will vary. One factor that was not tested in this experiment that is also important in sunglass design is the effect of different kinds of lens coatings and tints. This experiment also did not test the ability of the sunglasses to filter ultraviolet (UV) light. Finally, this experiment did not take into account people's personal preferences for different styles of sunglasses.

▶ Expanding Your Knowledge

1. Students can either bring in items from home or test various items that are in the lab. Have students make a bar graph that compares the light each surface reflects.

Additional Notes

LAB

13 TEACHER'S NOTES **ELECTRICITY**

Investigating How the Length of a Conductor Affects Resistance

MATERIALS (for each lab group)

- ✓ voltage and current probes (and adapter cables)
- ✓ CBL
- ✓ TI graphing calculator and black link cable
- ✓ shallow plastic tray about 30 cm long
- ✓ 2% NaCl solution
- ✓ metric ruler
- ✓ dual channel amplifier
- ✓ 6 V battery
- ✓ 3 wires with alligator clips on each end
- ✓ 2 iron nails
- ✓ graph paper (optional)

Students should wear safety goggles, gloves, and laboratory aprons while performing this experiment.

Remind students that wires that are conducting electricity can become very hot. Students should connect the battery only for short periods of time and immediately disconnect the battery after collecting data.

▶ Objectives

In this lab, students will

- ***measure*** the current and voltage of a circuit several times, changing the length of one of the conductors in the circuit each time.
- ***calculate*** the resistance of the circuit.
- ***make graphs*** of their data to determine how increasing the length of one of the conductors in a circuit affects the resistance of the circuit.

▶ Planning for the Experiment

Organizing your class

Having two to four students in each lab group works best.

Recommended time

1 lab period

Preparing the CBL System

- If possible, have a voltage and a current probe, a dual channel amplifier, a CBL, and a graphing calculator connected for each group ahead of time. The calculators should be loaded with the Vernier PHYSCI program.

Preparing other materials

- Make a 2 percent NaCl solution by dissolving 20 g of table salt (sodium chloride) in enough deionized water to make 1 L. Each group should fill its tray to a depth of about 0.5 cm.
- If possible, provide plastic metric rulers so that they can be placed in the bottoms of the trays to measure the distance between the nails.
- For best results, make sure the batteries you provide are either new or fully charged.
- Make sure that the iron nails you provide are long enough so that the heads of the nails can be connected to a wire while the tips of the nails are in the saltwater solution.

▶ Procedural Notes

General

- During the experiment, students may observe gas bubbles forming on one nail and the other nail oxidizing. You might want to discuss with them the way bleach is made commercially from salt water. This reaction is shown below.

$$\text{NaCl (dissolved)} \xrightarrow{\text{electric current}} \text{NaOCl (dissolved)} + H_2\text{ (gas)}$$

LAB 13 NOTES

The sodium chloride solutions can be washed down the drain with plenty of water, provided your school drains are connected to a sanitary sewer system with a treatment plant. Make sure that students wash their hands when they are done with the experiment.

Specifics

- **Step 10** When helping students assemble their circuits, make sure that the wires are making good contact everywhere in the circuit and that the nails are making good contact with both the wires and the salt water. Students might want to clip the nails to the plastic tray so that they stay upright.

▶ Sample Data

Circuit Data

Distance between nails (cm)	Voltage (V)	Current (A)	Resistance (Ω)
2	5.478	0.08536	64.18
4	5.506	0.07976	69.03
8	5.749	0.06577	87.41
12	5.562	0.06017	92.44
16	5.562	0.05621	98.95
20	5.462	0.05201	105.0
24	5.471	0.04758	115.0
28	5.562	0.04618	120.4

Answers

▶ Analyzing Your Results

1. Answers will vary. The calculation for a distance of 2 cm between the nails is shown below. Answers for all the distances are listed in the table above.

$$\text{resistance (2 cm)} = \frac{5.478 \text{ V}}{0.08536 \text{ A}} = 64.18\ \Omega$$

2. Graphs will vary. For the sample data provided:

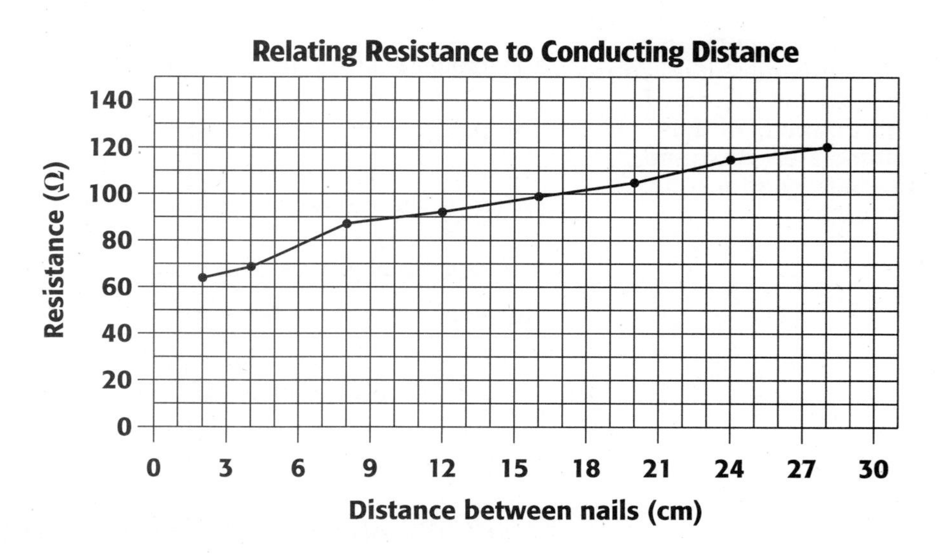

▶ Reaching Conclusions

3. As the conducting length of the salt water is increased (the distance between the two nails is increased), the resistance of the circuit also increases.
4. The wires conduct electricity more easily than the salt water. The Na^+ and Cl^- ions in the solution are able to spread out more than electrons that are confined to a thin wire. Also, the solution may be less conductive in some areas due to the ions' mobility.
5. The salt water has a greater resistance than the wires in the circuit for the same reasons discussed in item 4.
6. The maximum resistance per meter allowed to the house can be found by solving the following equations:

$$\text{maximum resistance} = \frac{110 \text{ V}}{30 \text{ A}} = 3.7\ \Omega$$

$$\text{maximum resistance per meter} = \frac{3.7\ \Omega}{45 \text{ m}} = 0.082\ \Omega/\text{m}$$

This means that the resistance per meter cannot be greater than 0.082 Ω/m.

▶ Defending Your Conclusions

7. Because wires are such good conductors, varying the length of one of them in the circuit would not produce a noticeable change in the resistance of the circuit.

▶ Expanding Your Knowledge

1. Students can consult a wire supplier or use the Internet to do their research. Their posters should summarize the advantages and disadvantages of different kinds of wires.

Additional Notes

LAB 13 NOTES

Testing Magnets for an Electric Motor

MATERIALS (for each lab group)
- ✓ magnetic field sensor (and adapter cable)
- ✓ CBL
- ✓ TI graphing calculator and black link cable
- ✓ bar magnet
- ✓ meterstick
- ✓ tape
- ✓ pencil
- ✓ graph paper (optional)

Students should wear safety goggles and gloves while performing this experiment.

If students are working in the laboratory, they should also wear laboratory aprons.

Make sure that students wash their hands when they are done with the experiment.

▶ Objectives

In this lab, students will
- ***measure*** the magnetic field strength of a bar magnet from several different distances.
- ***investigate*** how distance affects magnetic field strength.
- ***determine*** the distance from the magnet where the magnetic field is the strongest.

▶ Planning for the Experiment

Organizing your class

Having two to four students in each lab group works best.

Recommended time

1 lab period

Preparing the CBL System

- If possible, have a magnetic field sensor, a CBL, and a graphing calculator connected for each group ahead of time. The calculators should be loaded with the Vernier PHYSCI program.
- The magnetic field sensor is direction specific. It might be helpful the day before the lab to allow students to put the sensor in a magnetic field and then rotate the sensor so they can see how the field changes.

▶ Procedural Notes

General

- Caution students to handle the magnets carefully. If a magnet breaks, be sure to wear safety goggles and watch out for rough edges while carefully picking it up.
- Make sure students collect data so that the N and S poles are along the axis of the meterstick and the magnet is positioned so that the N pole is closer to the magnetic field sensor.
- Have students measure and record the length of the bar magnet they are using in centimeters so you can make better sense of their answers in the *Analyzing Your Results* section.

Specifics

- **Step 8** This lab's procedure says to use the LOW (GAUSS) setting. Depending on the magnet, students may need to use a more sensitive setting. Remind them to set the switch on the sensor accordingly.
- **Step 9** Because there are many sources of magnetic field interference in a classroom or a lab, remind students to get their sensors set up and secured in position before zeroing them.

▶ Sample Data

Magnetic Field Strength Data

Position of center of magnet (cm)	Magnetic field strength (G)	Position of center of magnet (cm)	Magnetic field strength (G)
0	1.95	15	– 2.01
1	2.00	16	– 3.15
2	2.01	17	– 3.76
3	3.15	18	– 7.04
4	3.76	19	– 9.05
5	8.95	20	– 17.98
6	16.45	21	– 18.56
7	18.56	22	– 16.45
8	17.98	23	– 8.95
9	9.05	24	– 3.76
10	7.04	25	– 3.15
11	3.66	26	– 2.01
12	3.25	27	– 2.04
13	2.12	28	– 1.93
14	1.98	29	– 1.87

Answers

▶ Analyzing Your Results

1. Graphs will vary. For the sample data provided:

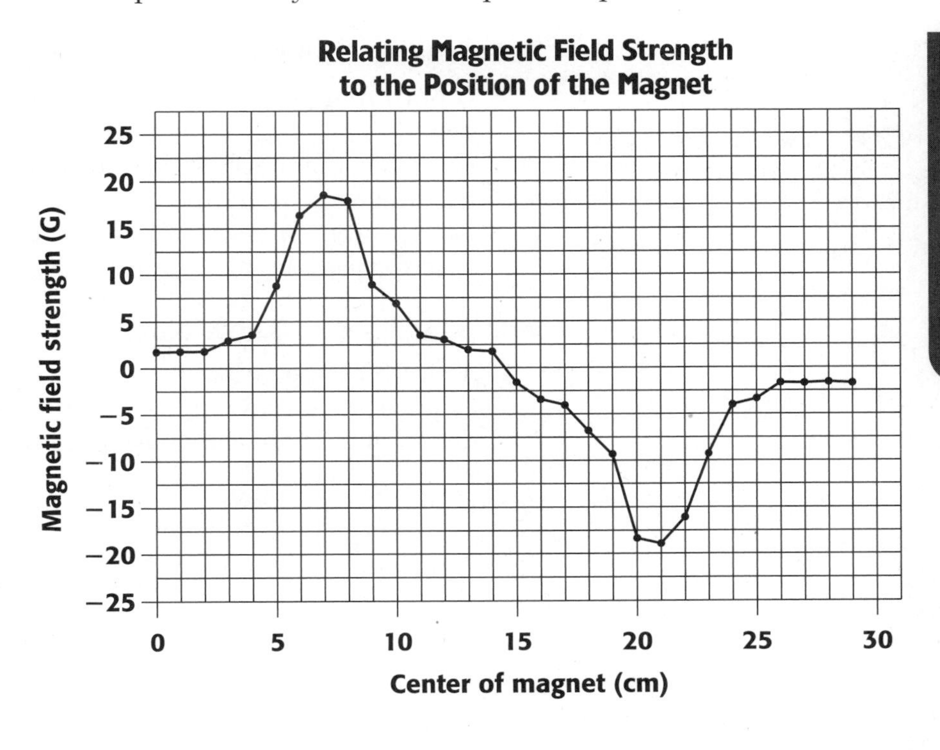

2. The magnetic field strength is low at first and then increases dramatically when the end of the magnet is within 1–2 cm of the magnetic field sensor (for the sample data provided, when the center of the magnet is positioned at about 7 cm). The magnetic field strength then gradually decreases until it becomes zero when the magnetic field sensor is halfway between the two poles (when the center of the magnet is positioned at about 14 cm). The trend for the S pole of the magnet is similar to the N pole, except that the magnetic field strength is negative.
3. The center of the magnet is positioned in the middle of the magnetic field sensor (at about 14 cm) when the magnetic field strength equals zero. The magnetic field strength is zero because the sensor is an equal distance from both the N pole and the S pole.
4. The graph would be the mirror image of the graph shown on the previous page, with the negative trough first and the positive peak second.

▶ Reaching Conclusions

5. Answers will vary. For the sample data provided, the center of the magnet is positioned at about 7 cm when the magnetic field is the strongest. The end of the magnet is within 1 cm of the sensor.
6. Answers will vary. For the sample data provided, the ends of the magnets could be spaced up to 2 cm apart because each magnet has a strong magnetic field that extends out about 1 cm.

▶ Defending Your Conclusions

7. If both magnets are the same, only one magnet needs to be tested for its magnetic field strength. You can assume that the other magnet behaves similarly. The distance apart that the magnets should be placed can then be determined by doubling the distance found to have the strongest magnetic field for the magnet you tested.

▶ Expanding Your Knowledge

1. A similar experiment can be performed using a solenoid or a coil of wire and a metal core. Make sure you approve students' plans before letting them perform the experiment.

Transmitting and Receiving a Message Using a Binary Code

MATERIALS (for each lab group)

- ✓ light sensor
- ✓ CBL
- ✓ TI graphing calculator and black link cable
- ✓ TI-Graph Link cable (only one needed)
- ✓ computer with a printer (only one needed)
- ✓ flashlight
- ✓ piece of cardboard

Students should wear safety goggles while performing this experiment.

If students are working in the laboratory, they should also wear gloves and laboratory aprons.

Make sure that students wash their hands when they are done with the experiment.

▶ Objectives

In this lab, students will

- ***create*** a binary code, and use the code to write a short message.
- ***transmit*** the message using a flashlight.
- ***decipher*** the message that is sent.
- ***evaluate*** the advantages and disadvantages of using a binary code to send a message.

▶ Planning for the Experiment

Organizing your class

Having four students in each lab group works best. All members of the group can help develop the code. Then part of the group will create and send a message, while the rest of the group deciphers it.

Recommended time

1 lab period

Preparing the CBL System

- If possible, have a light sensor, a CBL, and a graphing calculator connected for each group ahead of time. The calculators should be loaded with the Vernier PHYSCI program.

Preparing other materials

- Provide flashlights with focused beams of light. They permit the flashlight to be farther away from the light sensor.
- The graphs generated by the graphing calculators are much easier to decipher if they can be printed. To print the graphs, you need a TI-Graph Link cable that connects the calculator to the modem or printer port of your computer and the supporting software. You may want to review how to use the TI-Graph Link cable with your students prior to the lab.

▶ Procedural Notes

General

- Turning the overhead lights off during the experiment will give students better results.

Specifics

- **Step 2** Encourage students to write out very short messages rather than long ones. Also remind students that their messages should not be offensive to anyone.

- **Step 6** The memory of a TI-82 calculator is limited to storing 99 data points. If the light is sampled every 0.05 s, the sampling time is limited to 5 s. This should be adequate time for students to send the signal for one symbol. One way to extend the time is to lengthen the time between samples from 0.05 s to 0.1 s or 0.2 s. However, when the sampling time is increased, the graph loses some resolution and the results may be more difficult to decipher. Newer calculators, like the TI-83 and TI-83 Plus, will have enough memory to store about 300 data points, which will allow about 15 s to send a signal for one symbol.
- **Step 8** Make sure the flashlight will not roll off the table. Students might want to anchor the flashlight by putting it between some books.
- **Step 9** Emphasize to students that they should be sure to move the cardboard so that the light is either fully covered or not covered at all.

▶ Sample Data

Signal 1

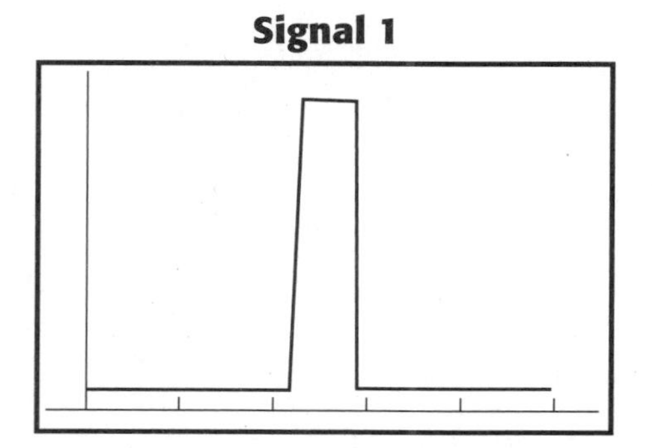

Signal 2

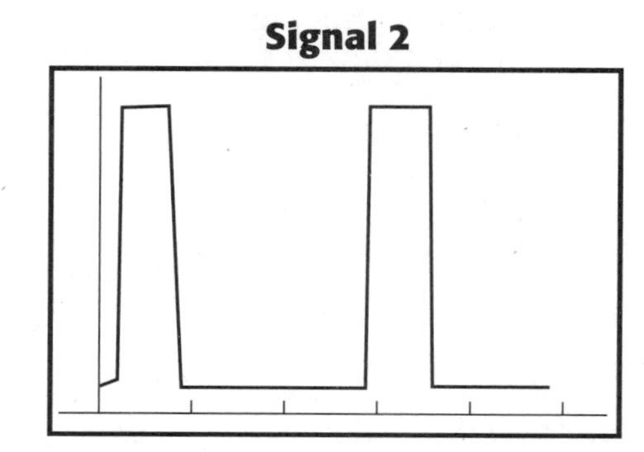

Answers

▶ Analyzing Your Results

1. Answers will vary. For the sample data provided, the graph of signal 1 would be translated as 00100. The graph of signal 2 would be translated as 10010.
2. Answers will vary for each group.

▶ Reaching Conclusions

3. Answers will vary. A few examples of other codes include Morse code, computer programming languages, pig Latin, flag signals, smoke signals, and hieroglyphics. There are several advantages to using binary codes. They are easily processed and interpreted by transducers and work well in electrical circuits because they can be represented by simple on-off switches. In addition, messages can be sent quickly. The main disadvantage of using a binary code is that there are only two values (1's and 0's), causing the number of bits needed to send a message to be very large sometimes.

4. Answers will vary. There are several ways to modify the code. One way is to insert an additional signal before each letter that is to be capitalized. In addition, each punctuation mark could have a unique signal.

5. $$\text{number of bits} = 1\ \cancel{\text{min}} \times \frac{60\ \cancel{\text{s}}}{1\ \cancel{\text{min}}} \times \frac{1\ \text{bit}}{0.000\ 164\ \cancel{\text{s}}} = 3.66 \times 10^5\ \text{bits}$$

   The answer for the next calculation will vary, depending on the number of bits each group used to represent each symbol. The calculation shown below is for five bits representing each symbol.

$$\text{number of symbols} = 3.66 \times 10^5\ \cancel{\text{bits}} \times \frac{1\ \text{symbol}}{5\ \cancel{\text{bits}}} = 73\ 200\ \text{symbols}$$

6. You could send an analog signal using a light source by varying the intensity of the light instead of turning it on and off. The graph of an analog signal would be smooth and continuous.

▶ Defending Your Conclusions

7. Answers will vary but might include light patterns and musical patterns. One example would be to transmit the frequencies of the three primary colors of light. Another example would be to transmit a short melody.

▶ Expanding Your Knowledge

1. ASCII code information can be found in some applied mathematics books as well as on the Internet or in computer reference books. A binary code including the numbers 65–90 represents the uppercase letters A–Z, while a binary code including the numbers 97–122 represents the lowercase letters a–z.

2. Both SETI and SERENDIP as well as several other organizations maintain interesting educational Web sites. On the sites, students will find explanations of technology as well as information about search techniques.

Additional Notes

Determining the Speed of an Orbiting Moon

MATERIALS (for each lab group)

- ✓ force sensor (and adapter cable)
- ✓ CBL
- ✓ TI graphing calculator and black link cable
- ✓ 200–300 g mass
- ✓ stopwatch
- ✓ 25 lb fishing line that is 1.0 m long
- ✓ swivel snap
- ✓ 0.5 in. PVC pipe 15 cm long
- ✓ meterstick
- ✓ marker
- ✓ graph paper (optional)
- ✓ calculator (optional)

Students should wear safety goggles and gloves while performing this experiment.

If students are working in the laboratory, they should also wear laboratory aprons.

This experiment involves students whirling a small mass around in circles over their heads. Make sure that students have enough room to collect data without hitting anyone or anything.

Make sure that students wash their hands when they are done with the experiment.

▶ Objectives

In this lab, students will

- ***make several models*** of a moon orbiting a planet, each time modeling a planet of a different mass.
- ***determine*** the distance each "moon" travels in 20 revolutions.
- ***calculate*** the speed of each "moon," and relate it to the mass of the planet being modeled.

▶ Planning for the Experiment

Organizing your class

Having at least three students in each lab group works best.

Recommended time

1–2 lab periods

Preparing the CBL System

- If possible, have a force sensor, a CBL, and a graphing calculator connected for each group ahead of time. The calculators should be loaded with the Vernier PHYSCI program.
- This lab's procedure is written for a dual-range force sensor, which can measure forces up to 50 N. If student force sensors are used, the CBL System will need to be set up differently, as described in the *Procedural Notes* section. In addition, students will not be able to do the last four trials because the student force sensor can't measure forces greater than 10 N.

Preparing other materials

- Use 25 lb nylon monofilament fishing line that is no longer than 1.0 m. Be sure to test the line to make sure it is strong enough to support the whirling mass.
- The mass should be between 200–300 g. To minimize the risk of injury, prepare the masses by enclosing 200–300 g of sand in cloth sacks or sandwich bags.
- You can buy swivel snaps at a fishing tackle store. Using swivel snaps will prevent the fishing line from twisting as the mass is whirled.
- Use small PVC pipes for the plastic tubes. Make sure that the plastic tubes are not rough or cracked at the edges so they do not fray the lines. You might have to sand the tubes to make them smooth with fine-grit sandpaper.

▶ Procedural Notes

General

- Try to have groups spaced as far apart as possible. Tell students to quickly warn others if a mass breaks loose.
- Remind students to stay in their own lab area and to be mindful of groups around them while they are collecting data.
- Do not allow students to swing the mass too excessively but just enough to maintain the necessary force reading.

Specifics

NOTE: Students may not be able to get a 0 reading on the force sensor. They should adjust the sensor to get as close to 0 as possible.

- **Steps 6 and 7** If students are using student force sensors, they should substitute the following steps for steps 6 and 7:
  6. Go to the MAIN MENU, and select SET UP PROBES. Enter "1" as the number of probes. Select FORCE from the SELECT PROBE menu. Enter "1" as the channel number.
  7. Select USE STORED from the CALIBRATION menu. Then select STUDENT FORCE from the FORCE PROBE menu.
- **Step 9** Make sure students do not let the force sensors slip or slide along the tabletop, as this will affect their readings.
- **Step 10** Caution students that before they start the mass swinging, they should make sure that the mass cannot slip through the knot, that the swivel snap is securely connected to the force sensor, and that the area around them is clear. When students finish timing, they should not stop the mass from swinging abruptly. Instead, they should gradually slow the mass down before stopping it completely.

▶ Sample Data

- Students may try to make the graph of their data fit a straight line. Remind them to be open to other possibilities. The actual relationship between the mass of a planet and the speed of its orbiting moon (for a given radius) is a square root function.

Forces and Orbiting Speeds

Force (N)	Time to complete 20 revolutions (s)	Distance traveled in 20 revolutions (m)	Orbiting speed (m/s)
5.0	21.00	62	3.0
7.5	17.59		3.5
10.0	15.50		4.0
12.5	13.39		4.6
15.0	12.14		5.1
17.5	11.39		5.4
20.0	10.99		5.6

Answers

▶ Analyzing Your Results

1. **distance traveled in one revolution = $2\pi \times 0.50$ m = 3.1 m**
   distance traveled in 20 revolutions = 20×3.1 m = 62 m

2. Answers will vary. The calculation for the orbiting speed when the force is 5.0 N is shown below. Answers for the speeds of all of the forces are listed in the table on the previous page.

$$\text{speed} = \frac{62 \text{ m}}{21.00 \text{ s}} = 3.0 \text{ m/s}$$

3. Graphs will vary. For the sample data provided:

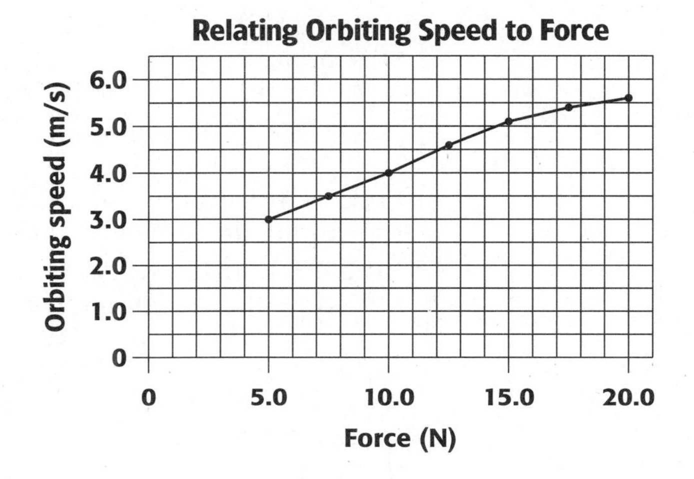

▶ Reaching Conclusions

4. The graph shows that as the force increases, the orbiting speed also increases.

5. The moon orbiting the planet that is twice as massive will move faster than the moon orbiting the less massive planet.

6. **mass of planet = $(320 \text{ m/s})^2 \times 250\,000\,000 \text{ m}$**

$$\times \left(1.499 \times 10^{10} \frac{\text{kg} \cdot \text{s}^2}{\text{m}^3}\right) = 3.8 \times 10^{23} \text{ kg}$$

▶ Defending Your Conclusions

7. Answers will vary. One way would be to repeat the experiment using a larger mass and greater forces.

▶ Expanding Your Knowledge

1. Supervise students as they perform the experiment again. This time, have students change the radius of the orbiting mass and keep the force constant. Students should find that the larger the radius, the slower the orbiting mass moves.

Relating Convection to the Movement of Tectonic Plates

MATERIALS (for each lab group)

- ✓ temperature probe
- ✓ CBL
- ✓ TI graphing calculator and black link cable
- ✓ aquarium or large tank
- ✓ small jar with a hole in the lid
- ✓ food coloring
- ✓ cold tap water
- ✓ hot tap water (you may need a hot plate and a metal pan)
- ✓ hot pad
- ✓ tongs
- ✓ stirring rod
- ✓ Teflon® tape
- ✓ stopwatch
- ✓ graph paper (optional)

Students should wear safety goggles, gloves, and laboratory aprons while performing this experiment.

Make sure students use a hot pad when handling hot water. Students should use tongs to transfer the jar filled with hot water into the aquarium.

▶ Objectives

In this lab, students will

- ***observe*** convection currents using an aquarium filled with cold water and a small jar filled with hot water and food coloring.
- ***measure*** temperature changes of four different areas in the aquarium over time.
- ***evaluate*** the theory that convection currents cause tectonic plates to move.

▶ Planning for the Experiment

Organizing your class

Having two to four students in each lab group works best.

Recommended time

1 lab period

Preparing the CBL System

- If possible, have a temperature probe, a CBL, and a graphing calculator connected for each group ahead of time. The calculators should be loaded with the Vernier PHYSCI program.

Preparing other materials

- If you do not have a supply of aquariums, large cooking pots will also work.
- Use small baby food jars to hold the heated water. You can make a hole in the center of the lids using a drill or a hammer and a nail. The hole needs to be about 3/8 in. in diameter.
- Use dark food coloring (blue or green) and very hot (near boiling) water.

▶ Procedural Notes

General

- Remind students to move the temperature probe slowly between stations so that convection currents are not disturbed.
- When students have recorded the temperature at station 4, they should leave the probe in the water and slowly return it to station 1. The probe should not be taken out of the water. This will reduce the amount of time needed for the probe to acclimate.
- Have a mop handy for cleaning up.

Specifics

- **Step 3** Provide large pitchers or beakers for students to fill the aquarium. Also make sure that the aquarium is where it should be before it is filled with water.

- **Step 11** Students should be advised to minimize the amount of air in the jar by filling the jar with hot water slowly. When students place the glass jar in the aquarium, air may get caught under the lid and prevent food coloring from leaving the jar. Students can insert the sharp end of a pencil into the hole to jiggle the air out. Once the air is out, the food coloring will rise steadily toward the surface.

▶ Sample Data

Temperature Readings and Observations

Time (min)	Temperature (°C)				Observations
	Station 1	Station 2	Station 3	Station 4	
0	16.32	16.56	16.44	16.44	None
0.5	16.71	17.49	19.69	20.42	Food coloring is rising steadily from the jar.
1	16.39	17.91	20.13	17.89	Food coloring is rising steadily from the jar in a single stream.
3	16.44	18.14	20.05	17.61	Food coloring is rising more slowly.
5	16.49	18.01	20.00	17.40	Food coloring is still rising but very slowly.
10	16.62	17.90	18.70	17.38	There is very little sign of convection movement in the tank.

The food coloring can be washed down the drain with plenty of water, provided your school drains are connected to a sanitary sewer system with a treatment plant. Make sure that students wash their hands when they are done with the experiment.

Answers

▶ Analyzing Your Results

1. Students may have difficulty with the concept of plotting data for all four stations on the same set of axes. You may wish to show students a sample graph. Graphs will vary. For the sample data provided:

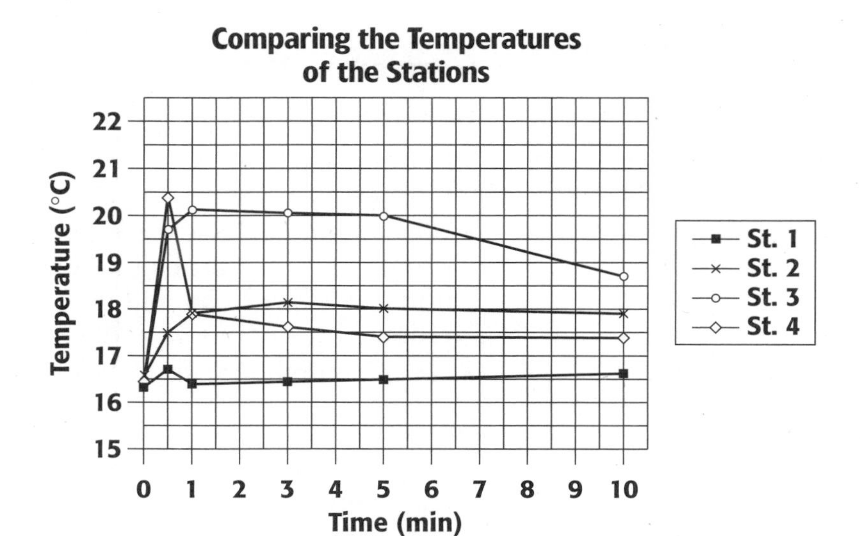

2. Station 3 showed the greatest temperature change over the 10-minute period. Station 1's temperature changed the least.
3. The graph shows that energy was transferred as heat from the jar to station 4, then to station 3, and finally to station 2.

▶ Reaching Conclusions

4. The food coloring left the jar more rapidly when the temperature differences were large. As the temperature differences declined, so did the rate of flow of food coloring from the jar.
5. Yes; when temperature differences are large (as they are in the asthenosphere due to residual heat from the formation of the Earth), hot material from the bottom of the tank moves upward and across the top of the tank.
6. This experiment predicts that lava will eventually stop rising in the asthenosphere. As a result, tectonic plates may eventually stop moving.

▶ Defending Your Conclusions

7. Answers will vary. Earth's asthenosphere consists of molten rock instead of water. The model used was also very small compared with the Earth, and energy as heat was transferred quickly. In addition, the model was very simplified. The movements in the Earth's asthenosphere are much more complex. A couple of ways to make the model more realistic would be to use a more viscous medium, like oatmeal, and to have a constant heat source at a much higher temperature in the bottom of the tank.

▶ Expanding Your Knowledge

1. You can divide students in groups to research topics, such as formation of the sun, formation of the inner planets, formation of the outer planets, asteroids, sources of heat in the solar system, and others. Each group should develop a poster with relevant images and information that can be presented to the class.

Additional Notes

Predicting Coastal Winds

MATERIALS (for each lab group)

- ✓ 2 temperature probes
- ✓ CBL
- ✓ TI graphing calculator and black link cable
- ✓ 150 mL beakers (2)
- ✓ tap water at room temperature
- ✓ sand
- ✓ reflector lamp with a clamp and a 300 W light bulb
- ✓ 2 ring stands
- ✓ 2 small clamps
- ✓ large rod

Students should wear safety goggles and gloves while performing this experiment.

If students are working in the laboratory, they should also wear laboratory aprons.

Make sure students do not touch the light source or look directly at it. Remind students to allow ample time for the lamp to cool before they handle it.

Make sure students handle glassware with caution and dispose of any broken glass in a specially designated container.

▶ Objectives

In this lab, students will

- ***make models*** of a coast on a sunny day using a lamp, sand, and water.
- ***compare*** the temperature of the sand with that of the water.
- ***predict*** which way the wind is likely to be blowing near the coast on a typical sunny day.

▶ Planning for the Experiment

Organizing your class

Having two to four students in each lab group works best.

Recommended time

1 lab period

Preparing the CBL System

- If possible, have the temperature probes, a CBL, and a graphing calculator connected for each group ahead of time. The calculators should be loaded with the Vernier PHYSCI program.
- This lab's procedure is written for two temperature probes to be used simultaneously. If each group has only one temperature probe, the lab can be run in two separate trials. The CBL System will have to be set up differently, as described in the *Procedural Notes* section.

Preparing other materials

- Give each group a hot pad or a padded glove in case they need to adjust the lamp while it is hot.
- Use dark sand or fine soil. Avoid wet soil or soil that has large spaces between the grains.
- A sun lamp works very well for this lab.

▶ Procedural Notes

General

- When students set up their apparatus, have them set it up away from the table edge. Warn students to be careful about getting entangled in the probes' cords and accidentally knocking the apparatus over. If possible, they should arrange the cords so that they are not in their work area.
- The temperature probes should be placed just below the surface of the water and the sand. If they are on the surface, the lamp will heat them directly, and the results will not be conclusive.

Make sure that students wash their hands when they are done with the experiment.

Specifics

- **Steps 2 and 4** If students are using only one temperature probe, they should substitute the following steps for steps 2 and 4:
  2. Plug the temperature probe into the Channel 1 input of the CBL. Connect the CBL to the graphing calculator by plugging the black link cable into the base of each unit.
  4. Go to the MAIN MENU, and select SET UP PROBES. Enter "1" as the number of probes. Select TEMPERATURE from the SELECT PROBE menu. Enter "1" as the channel number.
- **Step 7** When students are assembling the apparatus, make sure that there are no exposed electrical connections. If students have only one temperature probe and will do two separate trials, make sure that only the substance they are testing is under the lamp (not both substances, as the figure on page 74 shows).
- **Step 11** If students have only one temperature probe, they should turn the lamp off and allow it to cool for several minutes before testing the other substance.

▶ Sample Data

- These sample data were collected using a mixture of coarse sand and soil.

Temperature Changes of Water and Sand

Time (min)	Temperature of water (°C)	Temperature of sand (°C)
0	21.99	19.04
1	21.78	21.45
2	21.88	24.98
3	22.09	27.35
4	22.19	29.00
5	22.30	30.56
6	22.51	31.01
7	22.72	32.86
8	22.82	33.81
9	23.03	34.86
10	23.24	35.79
Total temperature change	1.25	16.75

Answers

▶ Analyzing Your Results

1. **Δt of water = 23.24°C − 21.99°C = 1.25°C**
   Δt of sand = 35.79°C − 19.04°C = 16.75°C

LAB 18 NOTES

▶ Reaching Conclusions

2. The sand had the greater temperature change during the 10-minute period.
3. Water has a very high specific heat, meaning it takes a lot of heat energy to raise the temperature of water even a little bit.
4. The winds would be blowing toward the shore in the middle of the day because the land (sand) would be very hot, causing the air above it to rise. The cooler air over the water would then move toward the shore.

▶ Defending Your Conclusions

5. Answers will vary. A real coastline differs from the model in many ways. Depending on the local terrain, winds may be affected by the presence of nearby inland lakes, coastal mountain ranges, and tall buildings in larger cities. In addition, the water, the soil, and the amount of sunlight varies from location to location. Therefore, the model does not allow you to predict exactly what will happen at a particular location. It can give only a general idea of what might happen.

▶ Expanding Your Knowledge

1. If students compare the wind direction with the presence of lows and highs, they may see some patterns.

Additional Notes

LAB

19 TEACHER'S NOTES USING NATURAL RESOURCES

Investigating the Effects of Acid Rain

MATERIALS (for each lab group)

- ✓ pH electrode (and adapter cable)
- ✓ CBL
- ✓ TI graphing calculator and black link cable
- ✓ deionized water
- ✓ buffer solution
- ✓ 2–3 samples from different bodies of water
- ✓ 100 mL beakers (5)
- ✓ rinse bottle filled with deionized water
- ✓ large beaker
- ✓ 0.10 M H_2SO_4 in a dropper bottle
- ✓ graph paper (optional)

Students should wear safety goggles, gloves, and laboratory aprons while performing this experiment.

Make sure you have Material Safety Data Sheets (MSDSs) available for all chemicals. Emphasize to students that sulfuric acid is corrosive and can cause burns if it comes in contact with skin. Caution students also not to breathe the acid's fumes. If a chemical comes in contact with a student's bare skin, have the student rinse the affected area immediately and continuously with lukewarm water.

▸ Objectives

In this lab, students will

- ***simulate*** the effects of acid rain by adding drops of acid to several different water samples.
- ***measure*** the pH of each water sample after adding each drop of acid.
- ***determine*** which water samples have the greatest buffering capacities.

▸ Planning for the Experiment

Organizing your class

Having two to four students in each lab group works best.

Recommended time

1 lab period

Preparing the CBL System

- If possible, have a pH electrode, a CBL, and a graphing calculator connected for each group ahead of time. The calculators should be loaded with the Vernier PHYSCI program.
- This lab's procedure suggests using the stored calibration values for the pH electrodes. Recalibrating the pH electrodes might give more accurate results. To check the calibration, you can use two standard buffer solutions of pH 4 and pH 7.

Preparing other materials

- You can purchase pH 7 buffer solution or you can make it by combining 500 mL of 0.10 M RA_2PO_4 with 291 mL of 0.10 M NaOH and diluting to 1L with deionized water. Check the pH and adjust to 7 if neccessary.
- You can purchase 0.10 M H_2SO_4 or you can make it by adding 5.6 mL of concentrated sulfuric acid to a 1 L volumetric flask containing about 500 mL of deionized water and then filling to the mark with deionized water. **Wear safety goggles, gloves, and a laboratory apron. Be sure you add acid to water and not water to acid. Add the acid with extreme caution, as its addition to water can generate heat and explosive spattering.** Pour small quantities of the solution into individual dropper bottles for each group.
- If hard water is not available in your area, you can dilute a buffer solution with a pH of 7 by a factor of 20 to mimic hard water.
- Fresh deionized water should be used, as carbon dioxide will dissolve in the water over time and lower its pH.
- Ideally, it is best if the water samples are collected right before students perform the experiment. Refrigerate the water samples if they must be stored.

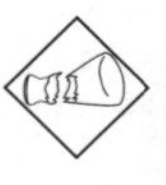

Make sure students handle glassware with caution and dispose of any broken glass in a specially designated container.

Follow the disposal instructions given in the *Procedural Notes* section. Make sure that students wash their hands when they are done with the experiment.

▶ Procedural Notes

General

- Emphasize that it is very important for students to label their samples clearly.
- The pH electrodes should not be allowed to dry out. Make sure students do not hit the sides or bottom of the beaker when they are testing the samples.
- If possible, obtain some water quality data on the waters you are testing. Important parameters include alkalinity, total dissolved solids (TDS), hardness, and productivity. Share any data you obtain with your students.
- Remind students to rinse the probe with deionized water between samples to avoid contamination.

Specifics

DISPOSAL INSTRUCTIONS

- Set out a disposal container that is clearly labeled for the *deionized water/acid waste.* Neutralize the solution with 0.10 M NaOH until the pH is between 5 and 9 before pouring it down the drain.
- The *buffer solution waste* can be washed down the drain with plenty of water, provided your school drains are connected to a sanitary sewer system with a treatment plant.
- Set out a disposal container that is clearly labeled for the *waste generated from the natural water samples.* Combine the solutions, adjust the pH if necessary with 0.10 M NaOH until it is between 5 and 9, and then wash it down the drain with plenty of water.

▶ Sample Data

- Sample 1 is from a brackish water lagoon.
- Sample 2 is from a productive golf course pond.

pH Data

Drops of acid added	Deionized water	Buffer solution	Sample 1	Sample 2
0	6.81	7.01	7.42	7.45
1	5.46	6.99	7.31	7.29
2	4.04	6.99	7.25	6.98
3	3.51	6.94	7.18	6.75
4	3.43	6.94	7.00	6.51
5	3.38	6.98	6.89	6.25
6	3.31	6.96	6.81	5.96
7	3.26	6.95	6.75	5.69
8	3.23	6.87	6.71	5.46
9	3.18	6.83	6.66	5.22
10	3.15	6.81	6.63	5.03

Answers

▶ Analyzing Your Results

1. Graphs will vary. For the sample data provided:

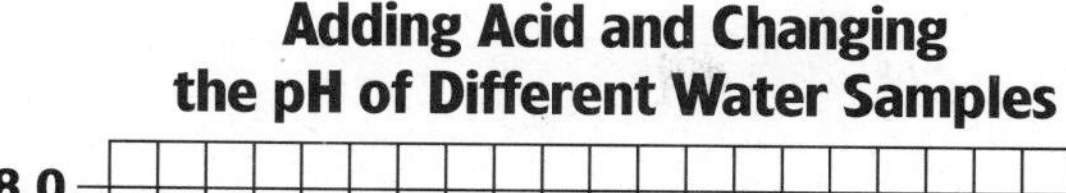

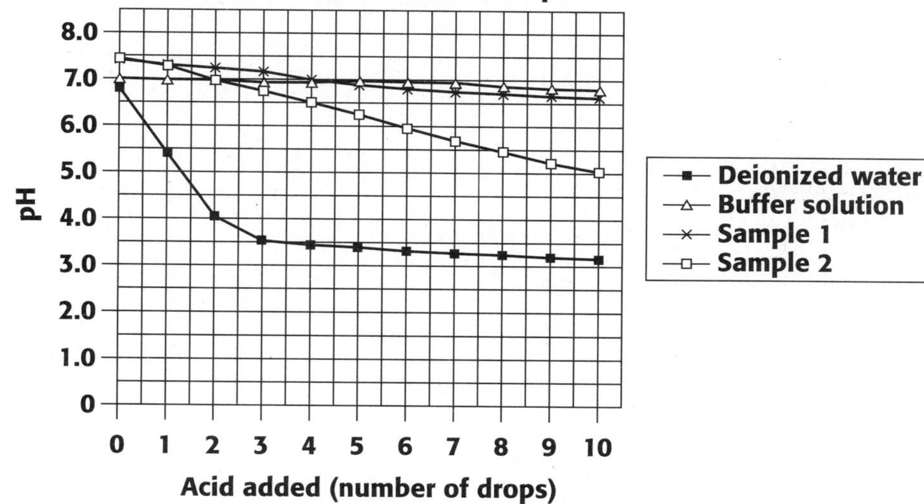

2. The pH of deionized water dropped considerably when acid was added, so it must have very little buffering capacity. The pH of the buffer solution changed only slightly, so it must have a high buffering capacity.

3. Answers will vary. For the sample data provided, the pH of sample 1 changed less than that of sample 2, so sample 1 has the greater buffering capacity.

▶ Reaching Conclusions

4. Deionized water has a low buffering capacity because it doesn't contain any dissolved salts to neutralize the acid.

5. Some waters have more dissolved salts that can react with acid and neutralize it.

6. If the organisms living in the body of water are sensitive to changes in pH, they might be adversely affected or even die.

7. Blood must be highly buffered. Its buffering capacity is probably most like that of the buffer solution.

▶ Defending Your Conclusions

8. Answers will vary but might include adding buffer solutions to natural waters may keep the pH from changing, but it might also create problems for the organisms living there.

▶ Expanding Your Knowledge

1. Have students compare the buffering capacities of buffered antacids with those that are not buffered.

Laboratory Experiments

This book was printed with acid-free recycled content paper, containing 10% POSTCONSUMER WASTE.

HOLT, RINEHART AND WINSTON

A Harcourt Classroom Education Company

Austin • New York • Orlando • Atlanta • San Francisco • Boston • Dallas • Toronto • London

Holt Science Spectrum: A Physical Approach

Laboratory Experiments

Author
E. David Thielk
Science Educator
Peninsula College
Port Townsend, WA

Reviewers
Rosa Hemphill
Chemistry and Research Teacher
Oregon Episcopal High School
Portland, OR

Bert Sherwood
Science Instructional Specialist
Socorro I.S.D.
El Paso, TX

Richard Sorensen
Vernier Software and Technology
Beaverton, OR

Safety Reviewer
Gregory Puskar
Laboratory Manager, Physics Department
West Virginia University
Morgantown, WV

All artwork on pages 2, 8, 19, 31, 36, 49, 50, 54, 57, 65, 70, and 74 created by Thomas Gagliano. All other artwork property of Holt, Rinehart and Winston.

Printed in the United States of America

ISBN 0-03-055579-5

2 3 4 5 6 862 03 02 01

Contents

Laboratory Experiments

CONTENTS, CONTINUED

WRITING A LABORATORY REPORT

In many of the laboratory investigations that you will be doing, you will be asking a question and then performing experiments to find out the answer to your question. Laboratory reports for these kinds of experimental investigations should contain the following parts:

▶ Title

This is the name of the laboratory investigation you are doing. If you are performing an investigation from this manual, the title of the report will be the same as the title of the investigation.

▶ Hypothesis

The hypothesis is what you think will happen during the investigation. It is often posed as an "If . . . then" statement. When you do an experiment, you will be changing one condition, or variable, and observing and measuring the effect of this change. The condition that you are changing is called the independent variable and should follow the "If" statement. The effect that you expect to observe is called the dependent variable and should follow the "then" statement.

The following statement is an example of a hypothesis: "If salamanders are reared in acidic water, then more of the salamanders will develop abnormally." The first part of this statement—"If salamanders are reared in acidic water"—is the independent variable because salamanders normally live in nearly neutral water, and you are changing this to acidic water. The second part of this statement—"then more of the salamanders will develop abnormally"—is the dependent variable. This is the change that you expect to observe and measure.

▶ Materials

This is a list of all the equipment and other supplies you will need to complete the investigation. If the investigation is taken from a laboratory manual, the materials are generally listed for you. In this manual, the materials are listed in a box on the left-hand side of the first page of the investigation.

▶ Procedure

The procedure is a step-by-step explanation of exactly what you did during the investigation. Investigations from this laboratory manual will have the procedures carefully written out for you. The procedure section should include an explanation of why your experiment was a controlled experiment. In a controlled experiment, you will change a single independent variable, and then compare the results with those obtained from a control group in which the independent variable was not changed.

For example, if you wish to see if acidic water increases the number of abnormalities among salamanders, your experiment must include a control group of salamanders raised under normal conditions so that you know what percentage of salamanders reared under normal conditions are abnormal.

▶ Data

Your data are your measurements and observations. They are often recorded in the form of tables, graphs, and drawings.

▶ Analyses and Conclusions

This part of the investigation explains what you have learned. You should evaluate your hypothesis and explain any errors you might have made during the investigation. Keep in mind that not all of the hypotheses you make will be correct. This is a normal part of being a scientist. You simply need to explain why things did not work out the way you thought they would. In this laboratory manual's investigations, there will be questions to guide you as you analyze your data. You should use these questions as a basis for your conclusions.

Additional Notes

SAFETY IN THE LABORATORY

Systematic, careful lab work is an essential part of any science program because lab work is the key to progress in science. In this class, you will practice some of the same fundamental laboratory procedures and techniques that scientists use to pursue new knowledge.

The equipment and apparatus you will use involve various safety hazards, just as they do for working scientists. You must be aware of these hazards. Your teacher will guide you in properly using the equipment and carrying out the experiments, but you must also take responsibility for your part in this process. With the active involvement of you and your teacher, these risks can be minimized so that working in the laboratory can be a safe, enjoyable process of discovery.

Anything can be dangerous if it is misused. Always follow the procedure for an experiment, and pay close attention to the safety precautions. Do not do anything differently unless told to do so by your teacher. If you follow the rules stated below, pay attention to your teacher's directions, and follow the cautions listed on chemical labels, equipment, and in the experiments, then you will stay safe.

▶ These safety rules always apply in the lab

1. **Wear safety goggles, gloves, and a laboratory apron.**
   Wear these safety devices whenever you are in the lab, not just when you are performing an experiment. Even if you aren't performing an experiment, laboratories contain chemicals that you should protect yourself from.

   If your safety goggles are uncomfortable or start to cloud up, ask your teacher for help. Try lengthening the strap, washing the goggles with soap and warm water, or using an anti-fog spray. Be sure to keep the strings of your laboratory apron tied at all times.

2. **No contact lenses in the lab.**
   Contact lenses should not be worn during any investigations using chemicals (even if you are wearing goggles). In the event of an accident, chemicals can get behind contact lenses and cause serious damage before the lenses can be removed. If your doctor requires that you wear contact lenses instead of glasses, you should wear eye-cup safety goggles in the lab. Ask your doctor or your teacher how to use this special eye protection.

3. **NEVER work alone in the lab.**
   Work in the lab only while under the supervision of your teacher. Do not leave equipment unattended while it is in operation.

4. **Wear the right clothing for lab work.**
   During the experiments, necklaces, neckties, dangling jewelry, long hair, and loose clothing can get caught in moving parts or catch on fire. Tuck in neckties or take them off. Do not wear a necklace or other dangling jewelry, including hanging earrings. It might also be a good idea to remove your wristwatch so that it is not damaged by a chemical spill. Wear shoes that will protect your feet from chemical spills and falling objects—no open-toed shoes or sandals, and no shoes with woven leather straps.

5. **Only books and notebooks needed for the experiment should be in the lab.**
   Only the lab notebook and perhaps the textbook should be used. Keep other books, backpacks, purses, and similar items in your desk or locker.

6. **Read the entire experiment before entering the lab.**
   Memorize the safety precautions. Be familiar with the procedure for the experiment. Only materials and equipment authorized by your teacher should be used. Your teacher will review any special safety precautions before the lab. If you are not sure of something, ask your teacher about it.

7. **Always heed safety symbols and cautions listed in the experiments, on handouts, and those posted in the room and given verbally by your teacher.**
   They are provided for a reason: YOUR SAFETY.

8. **Read chemical labels.**
   Follow the instructions and safety precautions listed on the labels.

9. **Be alert and walk with care in the lab.**
   Sometimes you will have to carry chemicals from the supply station to your lab station. Avoid bumping into other students and spilling the chemicals. Stay at your lab station at other times. Be aware of others that are near you or your equipment when you are about to do something. If you are not sure of how to proceed, ask.

10. **Know the proper fire drill procedures and the locations of fire exits and emergency equipment.**
    Make sure you know the procedures to follow in case of a fire or emergency.

11. **Know the locations of the safety showers and eyewash stations and how to use them.**

12. **If your clothing catches on fire, WALK (do not run) to the safety shower, stand under it, and turn it on.**
    Call your teacher while you do this.

13. **If you get a chemical in your eyes, walk immediately to the eyewash station, turn it on, and lower your head so that your eyes contact the running water.**
    Hold your eyelids open with your thumbs and fingers, and roll your eyeballs around. You must flush your eyes continuously for at least 15 minutes. Inform your teacher as you do this.

14. **If you spill a chemical on your skin, wash it off with lukewarm water, and call your teacher.**
    If you spill a solid chemical on your clothing, brush it off carefully without scattering it on somebody else, and call your teacher. If you get a liquid on your clothing, wash it off right away using the sink faucet, and call your teacher. If the spill is on your pants or something else that will not fit under the sink faucet, use the safety shower. Remove the pants or other affected clothing while you are in the shower, and call your teacher. (It may be temporarily embarrassing to remove pants or other clothing in front of your class, but failing to flush the chemical off your skin could cause permanent damage.)

15. **Report all accidents to your teacher IMMEDIATELY, no matter how minor.**
    In addition, if you get a headache, feel sick to your stomach, or feel dizzy, tell your teacher immediately.

16. **Report all spills to your teacher immediately.**
    Call your teacher rather than trying to clean a spill yourself. Your teacher will tell you if it is safe for you to clean up the spill; if not, your teacher will know how the spill can be cleaned up safely.

17. **The best way to prevent an accident is to stop it before it happens.**
    If you almost have an accident, tell your teacher so that you and your teacher can find a way to prevent it from happening again. Otherwise, the next time it could be a harmful accident instead of just a close call.

18. **Student-designed inquiry investigations, such as the Expanding Your Knowledge labs in this manual, must be approved by the teacher before being attempted by the student.**

19. **DO NOT perform unauthorized experiments or use equipment and apparatus in a manner for which they were not intended.**
    Use only the materials and equipment listed in an experiment's materials list or authorized by your teacher. Steps in a procedure should only be performed as described unless your teacher gives you permission to do otherwise.

20. **Food, beverages, chewing gum, and tobacco products are NEVER permitted in the lab.**

21. **Take only the amount of chemicals that you need.** However, if you happen to take too much and have some left over, DO NOT put it back into the original container. Ask your teacher what to do with any leftover chemicals.

22. **NEVER taste chemicals. Do not touch chemicals or allow them to contact areas of bare skin.**

23. **Use a sparker to light a Bunsen burner.** Do not use matches. Be sure that all gas valves are turned off and that all hot plates are turned off and unplugged before you leave the lab.

24. **Use extreme caution when working with hot plates or other heating devices.** Keep your head, hands, hair, and clothing away from the flame or heating area, and turn the devices off when they are not in use. Remember that metal surfaces connected to the heated area will become hot by conduction. Remember also that many metal, ceramic, and glass items do not always look hot when they are hot. Allow all items to cool before you store them.

25. **Do not use electrical equipment that has frayed or twisted wires.**

26. **Be sure your hands are dry before using electrical equipment.** Before plugging an electrical cord into a socket, be sure the electrical equipment is turned OFF. When you are finished with the device, turn it off. Before you leave the lab, unplug the device, but be sure to turn it off FIRST.

27. **Do not let electrical cords dangle from work stations; dangling cords might cause electrical shocks or make someone trip.** The area under and around electrical equipment should also be dry; cords should not lie in puddles of spilled liquid.

28. **Horseplay and fooling around in the lab are very dangerous.** Laboratory equipment and apparatus are not toys; never play in the lab or use lab time or equipment for anything other than their intended purpose.

29. **Keep work areas and apparatus clean and neat.** Always clean up any clutter made during the course of lab work. Put away apparatus in an orderly manner, and report any damaged or missing items.

30. **Always wash your hands thoroughly with soap and water at the conclusion of each experiment.**

31. **Whether or not the lab procedures remind you, all of these rules apply all of the time.**

▶ Safety Symbols

The following safety symbols will appear in the laboratory experiments to emphasize important additional areas of caution. Learn what they represent so you can take the appropriate precautions. Remember that the safety symbols represent hazards that apply to a specific activity, but the numbered rules given on the previous pages always apply to all work in the laboratory.

EYE PROTECTION

- Wear safety goggles, and know where the eyewash station is located and how to use it.
- An object that is swinging can cause serious injury.
- Avoid looking directly at a light source, as this may cause permanent eye damage.

HAND SAFETY

- Wear latex or nitrile gloves to protect yourself from chemicals in the lab.
- Use a hot mitt to handle resistors, light sources, and other equipment that may be hot. Allow equipment to cool before handling and storing it.

CLOTHING PROTECTION

- Wear a laboratory apron to protect your clothing.
- Tie back long hair, secure loose clothing, and remove loose jewelry to prevent these items from getting caught in moving parts or coming in contact with chemicals or heat sources.

HEATING SAFETY

- When using a Bunsen burner or a hot plate, always wear safety goggles and a laboratory apron to protect your eyes and clothing. Tie back long hair, secure loose clothing, and remove loose jewelry.
- Never leave a Bunsen burner or a hot plate unattended while it is turned on.
- If your clothing catches on fire, walk to the safety shower, and use the shower to put out the fire.
- Wire coils may heat up rapidly during some experiments. If heating occurs, open the switch immediately, and handle the equipment with a hot mitt.
- Allow all equipment to cool before storing it.

CHEMICAL SAFETY

- Do not eat or drink anything in the lab. Never taste chemicals.
- If a chemical gets on your skin or clothing or in your eyes, rinse it immediately and continuously with plenty of lukewarm water, and alert your teacher.
- If a chemical is spilled, tell your teacher, but do not clean it up yourself unless your teacher says it is okay to do so.

ELECTRICAL SAFETY

- Never close a circuit until it has been approved by your teacher. Never rewire or adjust any element of a closed circuit.
- Never work with electricity near water; be sure the floor and all work surfaces are dry.
- If the pointer of any kind of meter moves off the scale, open the circuit immediately by opening the switch.
- Light bulbs or wires that are conducting electricity can become very hot.
- Do not work with any batteries, electrical devices, or magnets other than those provided by your teacher.

GLASSWARE

- If a thermometer breaks, notify your teacher **immediately.**
- Do not use glassware that is broken, chipped, or cracked. Always use tongs or a hot mitt to handle heated glassware and other equipment because these items may not look hot when they are hot. Allow the equipment to cool before storing it.
- If a piece of glassware breaks, do not pick it up with your bare hands. Use a dustpan, a brush, and heavy gloves to place broken glass in a specially designated disposal container.
- If a light bulb breaks, notify your teacher immediately. Do not remove a broken bulb from the socket.

EXPLOSION PRECAUTION

- Use flammable liquids only in small amounts.
- When working with flammable liquids, be sure that there are no lit Bunsen burners or any other heat sources present.

GAS PRECAUTION

- Do not inhale a chemical's fumes directly.

ANIMAL SAFETY

- Be very careful when handling living organisms. They are delicate and can be harmed easily.

WASTE DISPOSAL

- Dispose of broken glass in a container specifically provided for that purpose.
- Dispose of any chemical waste only as instructed by your teacher.

HYGIENIC CARE

- Keep your hands away from your face and mouth.
- Always wash your hands thoroughly when you are done with an experiment.

LAB 1 INTRODUCTION TO SCIENCE

Designing a Pendulum Clock

MATERIALS

- ✓ motion detector
- ✓ CBL
- ✓ TI graphing calculator and black link cable
- ✓ several pieces of string, each of a different length
- ✓ several different hooked masses
- ✓ ring stand or table support
- ✓ metric ruler
- ✓ balance
- ✓ graph paper (optional)

Wear safety goggles and gloves while performing this experiment.

If you are working in the laboratory, wear a laboratory apron.

Set up your pendulum so that it is not possible to hit anyone when the pendulum swings.

Swing the pendulum only when you are collecting data.

When you are done with the experiment, wash your hands thoroughly.

▶ Objectives

- ***Make a hypothesis*** explaining how you think changing one variable of a pendulum will affect how the pendulum swings.
- ***Conduct*** an experiment to test your hypothesis.
- ***Evaluate*** the precision and accuracy of a pendulum for measuring time.

▶ Identifying the Problem

Imagine that you have traveled back to a time before modern clocks existed. A scientist named Galileo Galilei (1564–1642) has asked you to help him with an experiment he is conducting on falling objects.

Galileo is planning to drop several objects from the top of a tall tower and measure the exact number of seconds it takes for each object to hit the ground. Imagine that you have been asked to design a pendulum clock for the experiment that can accurately measure time in fractions of a second.

▶ Finding Out More Information

A *pendulum* is any body that is free to swing under the influence of gravity. Galileo used a very simple pendulum as a timing device in many of his experiments—a mass suspended from the end of a string.

The amount of time it takes a pendulum to swing forward and then back (one cycle) is always the same if the pendulum stays the same. But if you change the pendulum in some way, the time it takes to complete one cycle, called a *period,* also changes.

▶ Coming Up with a Plan

To design a pendulum clock, you have to know what factors affect the period of a pendulum. Some of these factors are the length of the pendulum string, the mass at the end of the pendulum, and how far the pendulum is pulled back before it is allowed to swing. A good place to start is to observe a pendulum while it is swinging. Based on your observations, you can make a hypothesis about how changing one of these factors, or independent variables, might affect how fast the pendulum swings.

Then you can design and carry out an experiment that tests your hypothesis. The results of the experiment will either support your hypothesis or disprove it. Either way, you will gain a better understanding of pendulums.

▶ Performing the Experiment

Preparing for the experiment

1. Prepare a table in your lab report similar to the one shown on the next page to record your data.

Pendulum Data

Independent variable (include units)	Period of the pendulum (s)			
	Trial 1	Trial 2	Trial 3	Average

USE YOUR DATASHEET

Setting up the CBL System

2. Plug the motion detector into the Sonic input of the CBL. Connect the CBL to the graphing calculator by plugging the black link cable into the base of each unit.
3. Turn on both the CBL and the calculator. Press PRGM on the calculator, and select the PHYSCI program.
4. Go to the MAIN MENU, and select SET UP PROBES. Enter "1" as the number of probes. Select MOTION from the SELECT PROBE menu.
5. From the MAIN MENU, select COLLECT DATA. Select TIME GRAPH from the DATA COLLECTION menu.
6. Enter "0.05" as the time between samples in seconds. Enter "99" as the number of samples. Press ENTER, and then select USE TIME SETUP.

Setting up the pendulum and observing its motion

7. Securely tie one end of the string to a hooked mass and the other end to the support, as shown below. You can use any mass and any length of string as long as the pendulum is able to swing freely.
8. Set up the motion detector so that it is directed toward the pendulum and is about 0.5 m away, as shown.

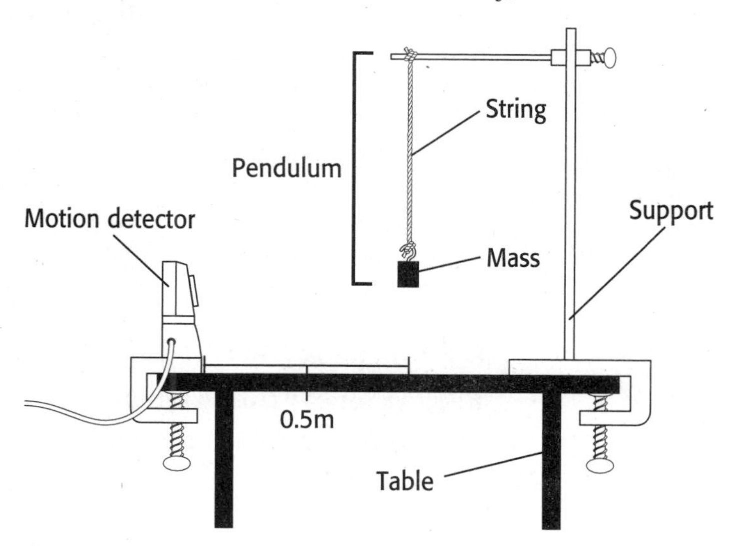

9. Pull back the mass, and press ENTER on the calculator. Release the mass, and let the pendulum swing freely.

10. When the CBL is done collecting data, stop the pendulum from swinging. Press ENTER and then select DISTANCE to view the graph, which should look similar to the one shown below. The time difference between two consecutive peaks is equal to the period of the pendulum.

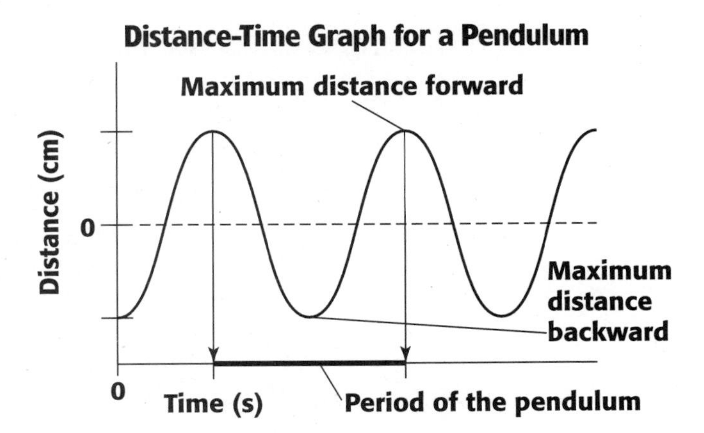

Making a hypothesis

11. Discuss with your lab group how changing each of the following might affect the period of the pendulum:
    a. the length of the pendulum string
    b. the mass at the end of the pendulum
    c. the distance between the pendulum and support when the pendulum is pulled back

12. Choose the one variable that you think will have the greatest effect on the period of the pendulum. Write a hypothesis that clearly states how changing this variable will affect the period of the pendulum. For example, if you increase your independent variable (length, mass, or distance), will it cause the period to be longer or shorter?

Designing and conducting the experiment

13. Do not change the pendulum you have already set up. Do one of the following, depending on what you chose as your independent variable:
    a. If you chose length, use a metric ruler to measure the length of the pendulum string.
    b. If you chose mass, unhook the mass from your pendulum, and use a balance to measure its mass.
    c. If you chose distance, pull the pendulum back as if you were about to let it swing. Hold it there, and use a metric ruler to measure the horizontal distance from the support to the bottom of the hooked mass.

    Record the measured value with the appropriate units in the first row of the column labeled *Independent variable* in your data table.

14. With the mass pulled back, press ENTER on the calculator. Then select RETURN and YES. Release the mass, and let the pendulum swing freely.

15. When the calculator has generated a graph of the motion, press ENTER and then select DISTANCE. Use the arrow keys to trace the graph to determine the period of the pendulum in seconds (the time difference between two adjacent peaks). Record the period in your data table.

16. Conduct two more trials without changing the pendulum by repeating steps 14 and 15 two more times. Be sure to treat the pendulum exactly the same each time. Record the period for each trial in your data table.

17. You will now change your independent variable by doing one of the following:
    a. If length is your independent variable, set up your pendulum using a string of a different length.
    b. If mass is your independent variable, set up your pendulum using a different hooked mass.
    c. If distance is your independent variable, pull your pendulum back so that the difference between the support and the bottom of the hooked mass when you start the pendulum swinging is different than it was before.

    Use the appropriate tool to measure the new independent variable, and record the value with the appropriate units in the next row of the column labeled *Independent variable* in your data table.

18. Repeat steps 14 and 15 three more times. Be sure to treat the pendulum exactly the same in all three trials. Record the period of the pendulum for each trial in your data table.

19. Change your independent variable once more. Use the appropriate tool to measure the independent variable, and then record the value in your data table. Repeat steps 14 and 15 three more times, making sure that you treat the pendulum exactly the same in all three trials. Record the period of the pendulum for each trial in your data table.

20. When you have finished, put away all of your materials, clean your work area, and wash your hands thoroughly.

▶ Analyzing Your Results

1. Determine the average period of the pendulum for each set of trials by using the following equation. Be sure your answers have the proper number of significant figures. Record your answers in your data table.

$$\textbf{average period} = \frac{\textbf{trial 1 period} + \textbf{trial 2 period} + \textbf{trial 3 period}}{\textbf{3}}$$

2. Plot your data on a graph like the one shown below. Plot the independent variable on the x-axis, being sure to include what your independent variable was and the units it was measured in. Plot the average period (the dependent variable) on the y-axis. If you use your graphing calculator, be sure to include a sketch of the graph in your lab report.

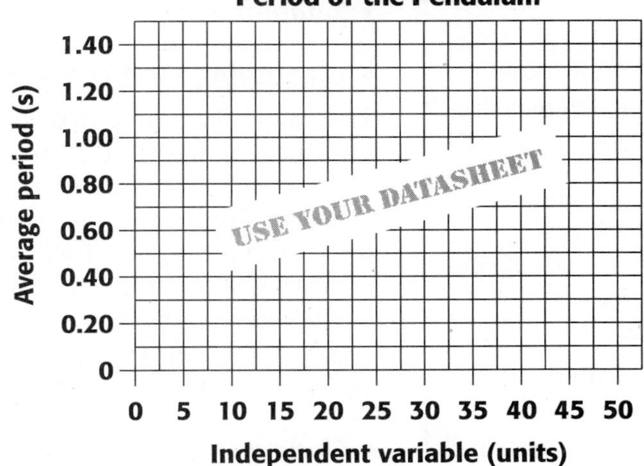

▸ Reaching Conclusions

3. Do your results support your hypothesis? Refer to your graph to explain why or why not.

4. Could the pendulum clock you made be adjusted to measure time accurately to the nearest one-thousandth (0.001) of a second? Explain why or why not.

▸ Defending Your Conclusions

5. Suppose someone tells you that if your results are precise, then they must be accurate. Is this a true statement? Explain why or why not.

▸ Expanding Your Knowledge

1. Research several different devices used to measure time, and compare the accuracy and precision of the devices.

2. Design a pendulum clock that can accurately measure 1 minute. Compete with your classmates to see who can develop the most accurate clock.

LAB 2 MATTER

Comparing the Buoyancy of Different Objects

MATERIALS

- ✓ force sensor (and adapter cable)
- ✓ CBL
- ✓ TI graphing calculator and black link cable
- ✓ 2 jars of different sizes, each with a small wire loop attached
- ✓ large container or laboratory sink
- ✓ 2 L graduated cylinder
- ✓ tap water
- ✓ sand
- ✓ piece of sturdy string
- ✓ scoop
- ✓ graph paper (optional)

Wear safety goggles and gloves while performing this experiment.

If you are working in the laboratory, wear a laboratory apron.

If a piece of glassware breaks, do not pick it up with your bare hands. Place broken glass in a specially designated disposal container.

When you are done with the experiment, wash your hands thoroughly.

▶ Objectives

- ***Measure*** the force exerted on an object in air and underwater.
- ***Calculate*** the buoyant force water exerts on the object.
- ***Relate*** the volume and mass of the object to its buoyancy.

▶ Identifying the Problem

The owner of a fish farm wants to divide a tank by suspending a net in the water. Some small weights will have to be attached to the bottom of the net so that it hangs straight down. In addition, 200 floats will have to be attached to the top of the net, each one providing a buoyant force of 5.0 newtons, or 5.0 N (1.1 lb). These floats will keep the top of the net at the surface of the water.

The owner wants to order floats that are just large enough to support the net. He has asked you to determine how large the floats must be to exert the necessary buoyant force.

▶ Finding Out More Information

When an object is submerged in water, there are two forces acting on it—the force of gravity and the buoyant force. Gravity pulls the object toward the center of Earth. The buoyant force is exerted by the water and acts in the opposite direction, exerting an upward push. If the buoyant force is greater than the force of gravity, the object will float. If the buoyant force is less than the force of gravity, the object will sink.

▶ Coming Up with a Plan

To determine the buoyant force acting on an object, you need to make two measurements with the force sensor. First you need to measure the force acting on the object in air, or the object's weight. You also need to measure the force acting on the object when it is submerged in water. The difference between these two forces is equal to the buoyant force acting on the object.

To determine how an object's volume affects the buoyant force acting on it, you can test jars of different volumes filled with sand. And if you change the mass of each jar by removing sand, you can determine how the mass of an object affects the buoyant force acting on it.

▶ Performing the Experiment

Preparing for the experiment

1. Prepare tables in your lab report similar to the ones shown on the next page to record your data.
2. Add enough tap water to the 2 L graduated cylinder so that each jar will be covered with water when placed inside. Also fill the large container or sink with tap water.

Table 1 Jar Volumes

Jar	Volume of water (mL)	Volume of water and jar (mL)	Volume of jar (mL)
1			
2		USE YOUR DATASHEET	

Table 2 Forces Acting on the Jars

Jar	Trial	Force acting on jar in air (N)	Force acting on jar underwater (N)	Buoyant force acting on jar (N)	Average buoyant force (N)
1	1				
	2		USE YOUR DATASHEET		
	3				
2	1				
	2				
	3				

Setting up the CBL System

3. Plug the force sensor into the Channel 1 input of the CBL. Connect the CBL to the graphing calculator by plugging the black link cable into the base of each unit.

4. Turn on both the CBL and the calculator. Press PRGM on the calculator, and select the PHYSCI program.

5. Go to the MAIN MENU, and select SET UP PROBES. Enter "1" as the number of probes. Select FORCE from the SELECT PROBE menu. Enter "1" as the channel number.

6. Select USE STORED from the CALIBRATION menu. Select DUAL RANGE from the FORCE PROBE menu, then select DUAL-RANGE 10 N. Set the switch on the sensor to 10 N.

7. From the MAIN MENU, select COLLECT DATA. Select MONITOR INPUT from the DATA COLLECTION menu. Force readings will be displayed on the calculator.

Determining the volumes of the jars

8. Record the initial volume of water in the graduated cylinder in milliliters in **Table 1.**

9. Fill both jars with sand and close them. Tie one end of the string to the loop attached to one of the jars. Hold the other end of the string so that the jar is suspended in air.

10. Slowly and carefully lower the jar into the graduated cylinder until the jar is resting on the bottom. Record the combined volume of the water and jar in **Table 1.** Then remove the jar from the graduated cylinder.

11. Repeat steps 8–10 for the second jar.

Measuring forces

12. Away from the container of water, tie the end of the string you were holding to the loop on the force sensor. Holding the force sensor by the handle, suspend the jar in air from the force sensor. When the reading stabilizes, record the force acting on the jar in air in newtons in **Table 2.**

13. Leaving the jar suspended from the force sensor, lower the jar into the container of water, as shown below. Make sure that the jar does not touch the bottom or sides of the container. (You might want to stabilize the force sensor with a ring stand.) When the reading stabilizes, record the force acting on the jar underwater in **Table 2.** Then remove the jar from the water.

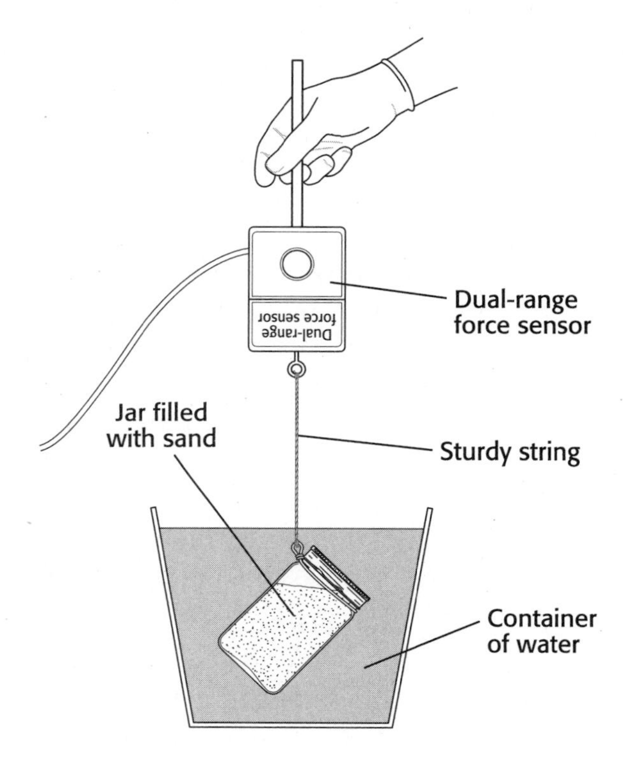

Determining how mass affects buoyancy

14. Remove several scoops of sand from the jar, and repeat steps 12 and 13. Record your data in **Table 2.**

15. Repeat step 14 once more.

16. When you have finished testing the first jar, remove it from the string and attach the second jar. Repeat steps 12–15 for this jar.

17. When you have finished, press "+" on the calculator. Then put away all of your materials, clean your work area, and wash your hands thoroughly.

▶ Analyzing Your Results

1. Calculate the volume of each jar by using the following equation. Record your answers in **Table 1.**

volume of jar = volume of water and jar − volume of water

2. Calculate the buoyant force for each jar for each trial by using the following equation. Record your answers in **Table 2.**

buoyant force acting on jar = force acting in air − force acting in water

3. Calculate the average buoyant force for each jar by adding the three buoyant forces and dividing by 3. Record your answers in **Table 2.**

4. Obtain data from another group. Plot your data and their data on a graph like the one shown below. Connect the data points with the line or smooth curve that fits the points best. Be sure to label your graph clearly.

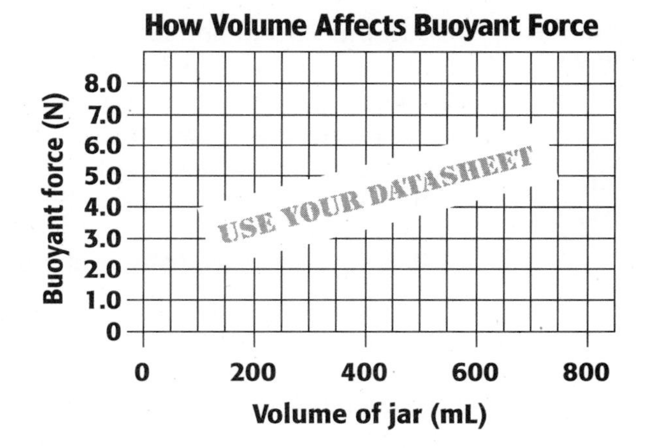

▶ Reaching Conclusions

5. How is an object's buoyancy affected by its volume? How does an object's mass affect its buoyancy?

6. Use your graph to determine what size float the owner should buy to provide 5.0 N of buoyant force.

▶ Defending Your Conclusions

7. Many floating objects are less dense than water. But boats are made of materials that are denser than water. What can you conclude about the buoyant force acting on a boat compared with the force of gravity acting on the boat?

▶ Expanding Your Knowledge

1. Design your own boat using thick aluminum foil. Compete with your classmates to see whose boat can carry the most pennies before sinking.

CHAPTER 2

Predicting the Physical and Chemical Properties of Elements

MATERIALS

✓ TI graphing calculator (or computer with graphing software or graph paper and a pencil)

Wear safety goggles while performing this experiment.

If you are working in the laboratory, wear gloves and a laboratory apron.

When you are done with the experiment, wash your hands thoroughly.

▶ Objectives

- ***Make graphs*** for some of the physical and chemical properties of the first 20 elements in the periodic table.
- ***Observe patterns*** in the properties of elements based on their positions in the periodic table.
- ***Predict*** the properties of two elements.

▶ Identifying the Problem

Imagine that you are a chemist in 1869. The recent development of the periodic table by the Russian chemist Dmitri Mendeleev has revolutionized the science of chemistry. More than 60 elements have been identified, and many of their physical and chemical properties have been determined.

Imagine that in your own laboratory, you have collected data for all but two of the first 20 elements. You still have to collect data for lithium (the 3rd element) and for fluorine (the 9th element). You notice as you analyze your data that there are some patterns, so you decide to make a graph of your data. That way you can use these patterns to predict the properties of the two elements you haven't collected data for yet.

▶ Finding Out More Information

When Mendeleev published the first periodic table of the elements in 1869, the response from chemists throughout Europe was very positive. Mendeleev saw that the properties of elements followed a pattern. In his periodic table, Mendeleev sometimes found it necessary to leave gaps so that elements with similar properties would line up in the same group, or column. As new elements were discovered, these gaps eventually filled. But even before these "missing" elements were discovered, Mendeleev predicted their properties based on their prospective positions in the periodic table.

▶ Coming Up with a Plan

The patterns for the properties of the first 20 elements are not obvious when you first look at the data in the table on the next page. But if you plot the atomic number of the element on the x-axis and a property of the element on the y-axis, the pattern for that property becomes obvious. You can then use the graph to predict that property for lithium and for fluorine.

Because there are four different properties listed in the table on the next page, you can divide into groups of four. Each person in your group can make a graph for a different property. Then you can compare the graphs to see how the patterns are similar. Comparing all four graphs that are generated will allow you to group elements with similar properties.

Some Properties of the First 20 Elements

Element	Atomic number	Density (g/cm^3)	Melting point (°C)	Boiling point (°C)	Charge of ion*
H	1	0.088	−259	−253	1+
He	2	0.21	−272	−269	0
Li		USE YOUR DATASHEET			
Be	4	1.85	1283	2970	2+
B	5	2.34	2300	2550	0
C	6	2.25	3570	4200	0
N	7	1.00	−210	−196	3−
O	8	1.40	−218	−183	2−
F		USE YOUR DATASHEET			
Ne	10	1.40	−249	−246	0
Na	11	0.097	98	889	1+
Mg	12	1.74	650	1120	2+
Al	13	2.70	660	2327	3+
Si	14	2.33	1414	2355	0
P	15	1.83	44	280	3−
S	16	2.07	119	444	2−
Cl	17	2.40	−101	−34	1−
Ar	18	1.62	−189	−186	0
K	19	0.86	64	774	1+
Ca	20	1.55	845	1420	2+

* A zero indicates that an atom of that element does not form an ion.

▶ Performing the Experiment

Setting up the graphing calculator

1. Turn on the graphing calculator. Press STAT. Select CLRLIST from the EDIT menu. Press 2nd “L1,” then “,”. Then press 2nd “L2,” “,”, 2nd “L3,” and ENTER. This will clear lists L1–L3.

2. Press 2nd MEM. Select DELETE from the MEMORY menu. Select Y-Vars, then press ENTER to delete each Y-Var that is stored in the calculator’s memory.

3. Press STAT. Select EDIT from the EDIT menu. Label the first column “L1” and the second column “L2.” (They may already be labeled.)

Entering the data

4. In groups of four, decide who will plot each property. In the first column, enter the atomic numbers. Do this by pressing 1, then ENTER, 2, then ENTER, 4, then ENTER, etc. until you have entered all the atomic numbers listed in the table. Do not enter the numbers 3 and 9.

5. Use the right arrow key to move the cursor to the second column. Enter the data for the property you will be graphing in the same way. Ignore the data that is missing from the table.

Viewing the graph

6. Press 2nd STAT PLOT. Select PLOTSOFF, and then press ENTER. Press 2nd STAT PLOT again, then select PLOT1. Turn PLOT1 on by selecting ON and pressing ENTER. Use the arrow key to move the cursor to TYPE, and select the line graph (the second option). Press ENTER.
7. Enter "L1" for the Xlist and "L2" for the Ylist. Select the small square for the MARK.
8. Press ZOOM. Use the arrow key to scroll down to ZOOM-STAT, and press ENTER. You will see a graph of the data.
9. Press TRACE, and use the arrow keys to move across the graph. As you do, the x- and y-values will be displayed at the bottom of the screen.

▶ Analyzing Your Results

1. Use the arrow keys to move the cursor to the data points for the elements with atomic numbers 2 and 4. After studying the graph and reviewing the data in the table, predict the value of the property for the element with atomic number 3 (lithium). To which family does lithium belong?
2. Use the arrow keys to move the cursor to the data points for the elements with atomic numbers 8 and 10. Then predict the value of the property for the element with atomic number 9 (fluorine). To which family does fluorine belong?

▶ Reaching Conclusions

3. Are carbon (atomic number 6) and silicon (atomic number 14) in the same group of the periodic table? Compare the properties of these two elements.
4. Make a similar comparison for nitrogen (atomic number 7) and phosphorus (atomic number 15). Are these two elements in the same group?

▶ Defending Your Conclusions

5. Obtain actual data for lithium and for fluorine from your teacher. Compare your predictions with the actual values. How accurate were your predictions? Why might there be some variation from the actual values?

▶ Expanding Your Knowledge

1. Pick an element with an atomic number greater than 20, and research it. Find out when and how it was discovered, as well as who discovered it. Also find out its physical and chemical properties, its abundance on Earth, and any current uses it has in manufacturing or technology. Present your findings to the class.

LAB

4 THE STRUCTURE OF MATTER

Determining Which Household Solutions Conduct Electricity

MATERIALS

- ✓ conductivity probe (and adapter cable)
- ✓ CBL
- ✓ TI graphing calculator and black link cable
- ✓ table sugar solution
- ✓ table salt solution
- ✓ baking soda solution
- ✓ Epsom salt solution
- ✓ dishwashing detergent solution
- ✓ rubbing alcohol solution
- ✓ milk of magnesia solution
- ✓ chemical fertilizer solution
- ✓ 100 mL beakers (3)
- ✓ rinse bottle filled with deionized water
- ✓ large beaker
- ✓ tissue paper

Wear safety goggles, gloves, and a laboratory apron while performing this experiment.

Never eat or drink anything in the laboratory.

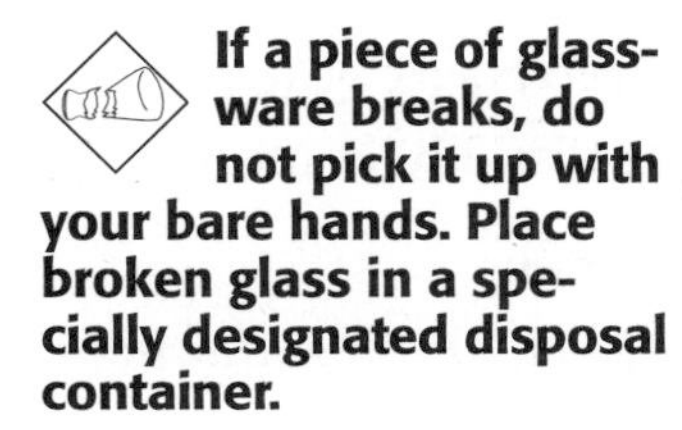

If a piece of glassware breaks, do not pick it up with your bare hands. Place broken glass in a specially designated disposal container.

▶ Objectives

- ***Measure*** the conductivities of several dilute samples of household solutions.
- ***Relate*** the conductivity of each sample to the kinds of bonds that are present in each substance.
- ***Determine*** which solutions are likely to be contributing to the increased conductivity of a river.

▶ Identifying the Problem

The water-quality department of a small city has recently discovered that the local river has a higher than normal conductivity. Although the city now requires that all new houses and office buildings be connected to a sewer treatment plant that can remove some dissolved solids, waste water has traditionally been released to the ground.

Unfortunately, this waste water (and whatever is dissolved in it) eventually seeps through the aquifer and enters the river. Imagine that you have been asked to investigate which household solutions in the waste water are likely to be contributing to the river's high conductivity and then to report your findings to the city council.

▶ Finding Out More Information

Ionic bonds are formed by the strong attraction between oppositely charged ions. When an ionic compound is dissolved in water, these bonds are broken, and the individual ions spread throughout the solution. The solution can conduct electricity because these charged ions can move easily. The more dissolved ions there are in the solution, the greater the solution's conductivity.

Covalent bonds, on the other hand, generally are not broken when dissolved in water. And each molecule has no overall charge. Therefore, covalently bonded compounds usually do not conduct electricity when they are dissolved in water.

▶ Coming Up with a Plan

One way to determine which household solutions are able to conduct electricity is to use a conductivity probe to test the solutions. A conductivity probe passes a small electric current through the solution and measures how easily the solution is able to conduct the current. The conductivity probe gives a numerical reading of the solution's conductivity in SI units of microsiemens per centimeter, or μS/cm. The higher the reading, the greater the conductivity of the solution. This means more ions must be present.

Rubbing alcohol is flammable, so make sure that no lit Bunsen burners or other heat sources are present in the laboratory during this experiment.

When you are done with the experiment, dispose of the waste as instructed by your teacher, and wash your hands thoroughly.

▶ Performing the Experiment

Preparing for the experiment

1. Prepare a table in your lab report similar to the one shown below to record your data.

Table 1 The Conductivity of Some Household Solutions

Beaker number	Household solution	Conductivity (μS/cm)	Ionic, covalent, or polyatomic?
1	Tap water		
2	Table sugar		
3	Table salt		
4	Rubbing alcohol		
5	Milk of magnesia		
6	Epsom salt		
7	Baking soda		
8	Dishwashing detergent		
9	Chemical fertilizer		

2. Obtain three 100 mL beakers, and label them 1, 2, and 3.
3. Obtain about 20 mL of each of the first three household solutions (as listed in **Table 1**) in the appropriate beakers.

Setting up the CBL System

4. Plug the conductivity probe into the Channel 1 input of the CBL. Connect the CBL to the graphing calculator by plugging the black link cable into the base of each unit. Set the switch on the probe amplifier box to 0–20 000 μS.
5. Turn on both the CBL and the calculator. Press PRGM on the calculator, and select the PHYSCI program.
6. Go to the MAIN MENU, and select SET UP PROBES. Enter "1" as the number of probes. Select CONDUCTIVITY from the SELECT PROBE menu, and enter "1" as the channel number.
7. Select USE STORED from the CALIBRATION menu. Select 0–20 000 MICS (μS) from the CONDUCTIVITY menu.
8. From the MAIN MENU, select COLLECT DATA. Select MONITOR INPUT from the DATA COLLECTION menu.

Measuring the conductivities of the samples

NOTE: *Be sure to take your readings from the calculator and not the CBL.*

9. Starting with the first beaker, submerge the tip of the probe in the liquid and gently swirl. Once the reading has stabilized, record the conductivity in **Table 1.**

10. Rinse the conductivity probe thoroughly with deionized water, using a large beaker to catch the rinse water. Then gently blot the probe dry with tissue paper.
11. Repeat steps 9 and 10 for solutions 2 and 3. Record each conductivity reading in **Table 1.**
12. After you have finished testing all three solutions, dispose of them as instructed by your teacher. Clean the beakers thoroughly, rinse them with deionized water, and then label them 4, 5, and 6.
13. Obtain about 20 mL each of solutions 4, 5, and 6 in the appropriate beakers. Repeat steps 9–11.
14. Repeat step 12, this time labeling the beakers 7, 8, and 9.
15. Obtain about 20 mL each of solutions 7, 8, and 9 in the appropriately numbered beakers. Repeat steps 9–11 once more.
16. When you have finished testing these solutions, dispose of them as instructed by your teacher and then press "+" on the calculator. Make sure you put away all of your materials, clean your work area, and wash your hands thoroughly before you leave.

▶ Analyzing Your Results

1. Make a table in your lab report similar to the one shown below. Complete the table by putting each solution you tested into one of the three categories.

Table 2 Classifying Household Solutions

No conductivity (0 μS/cm)	
Low conductivity (1–500 μS/cm)	USE YOUR DATASHEET
High conductivity (>500 μS/cm)	

▶ Reaching Conclusions

2. Which household solutions might be contributing to the high conductivity of the river? Explain why.
3. Explain why tap water shows some conductivity.
4. Which household solution has the highest conductivity?
5. Given the fact that there is a large golf course and several large gardens that have to be maintained near the river, which household item is probably the greatest contributor to the high conductivity of the river?

6. The table below lists the chemical formulas and some comments for the substances in each solution you tested. Use your data and this information to make a general conclusion about the ability of solutions of organic compounds to conduct electricity. Are there any exceptions to your general conclusion? Explain.

Chemical Formulas of Some Household Substances

Household substance	Chemical formula and additional comments
Tap water	H_2O with some dissolved minerals
Table sugar	$C_{12}H_{22}O_{11}$ (organic compound)
Table salt	NaCl
Rubbing alcohol	C_3H_8O organic compound dissolved in water
Milk of magnesia	$Mg(OH)_2$ suspended in water
Epsom salt	$MgSO_4$
Baking soda	$NaHCO_3$
Dishwashing detergent	Complex organic mixture; main ingredient is sodium lauryl sulfate, $C_{12}H_{25}SO_4Na$
Chemical fertilizer	Complex mixture of nutrients and minerals; includes nitrates, such as NO_3^-, and phosphates, such as PO_4^{3-}

7. For each solution you tested, use your data, your textbook, and the information provided in the table above to determine whether each substance has ionic or covalent bonds or contains one or more polyatomic ions. Record your answers in **Table 1.**

▶ Defending Your Conclusions

8. Write a brief recommendation to the city council with some suggestions about what homeowners and local businesses can do to reduce the conductivity of the river. Explain to them why following your recommendations would help reduce the river's conductivity.

▶ Expanding Your Knowledge

1. Collect samples from several water sources in your area, such as wells, lakes, rivers, streams, and bayous, and determine each sample's conductivity. Contact the local water-quality experts to find out what kinds of substances may be contributing to the conductivity of the water in your area.

2. Develop a plan to test the conductivity of soil. Compare the conductivities of several samples of soil from different locations.

LAB

5 CHEMICAL REACTIONS

Investigating the Effect of Temperature on the Rate of a Reaction

MATERIALS

- ✓ temperature probe
- ✓ CBL
- ✓ TI graphing calculator and black link cable
- ✓ water bath
- ✓ plastic test-tube rack
- ✓ 2 test tubes
- ✓ 10 mL graduated cylinders (2)
- ✓ warm tap water
- ✓ cold tap water
- ✓ deionized water
- ✓ 50 mL beakers (2)
- ✓ 1.0 M HCl
- ✓ 0.10 M $Na_2S_2O_3$
- ✓ stopwatch
- ✓ graph paper (optional)

There is a possibility that you may experience an allergic reaction to either sodium thiosulfate or sulfur. If you think you might be having a reaction, tell your teacher immediately.

Wear safety goggles, gloves, and a laboratory apron while performing this experiment.

Hydrochloric acid is corrosive. Avoid breathing the acid's fumes. If a chemical gets on your skin or clothing or in your eyes, rinse the affected area immediately with lukewarm water, and alert your teacher.

▸ Objectives

- ***Perform and observe a reaction*** several times, at a different temperature each time.
- ***Measure*** the time it takes for each reaction to be completed.
- ***Make a graph*** of your data.
- ***Determine*** how the rate of a chemical reaction varies with temperature.

▸ Identifying the Problem

Imagine that you are starting your own photographic film development business. To be successful, you will need to come up with a film development process that will allow you to develop film quickly while still maintaining a consistent quality.

You have noticed that the temperature of the chemicals in your processing lab seems to affect the time that it takes for one of the film-exposing reactions to occur. You have decided to investigate further by varying the temperature of the reactants to see exactly how this changes the rate of the reaction.

▸ Finding Out More Information

Sodium thiosulfate, $Na_2S_2O_3$, is a chemical used to develop film. In a step known as "fixing," thiosulfate ions, $S_2O_3^{2-}$, combine with unexposed silver bromide, AgBr, and remove it from the photographic paper. This process leaves exposed silver "fixed" on the paper, forming the images that are in the picture.

Because hydrochloric acid, HCl, combines with sodium thiosulfate in a similar reaction and is much less expensive than silver bromide, you have decided to use it instead of silver bromide for your experiment. The reaction you will carry out can be described by the following equation:

$$Na_2S_2O_3 \text{ (dissolved)} + 2HCl \text{ (dissolved)} \rightarrow S \text{ (solid)} + H_2SO_3 \text{ (dissolved)} + 2NaCl \text{ (dissolved)}$$

▸ Coming Up with a Plan

You can't measure the temperature of the reactants directly because you would contaminate the temperature probe. Instead, you can put the test tubes holding the reactants in a water bath and then measure the temperature of the water surrounding them. After several minutes, the temperature of the reactants should be about the same as that of the water.

You also need to know when the reaction is complete. Both reactants, as well as most of the products, are dissolved in water. But one product, sulfur, S, is a light-colored solid. Therefore, when you see a light-colored solid form, you can assume that the reaction is complete.

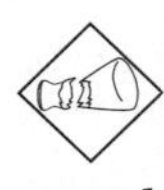

If a piece of glassware breaks, do not pick it up with your bare hands. Place broken glass in a specially designated disposal container.

When you are done with the experiment, dispose of the waste as instructed by your teacher, and wash your hands thoroughly.

▶ Performing the Experiment

Preparing for the experiment

1. Prepare a table in your lab report similar to the one shown below to record your data.

Temperature and Reaction Time of Reactants

Trial	Temperature (°C)	Elapsed reaction time (s)
1		
2		
3		
4		

2. Obtain two test tubes in a test-tube rack, two 10 mL graduated cylinders, and two 50 mL beakers. Label one of the beakers and one of the graduated cylinders "HCl." You will use this glassware to obtain and measure the hydrochloric acid solution. Label the other beaker and graduated cylinder "$Na_2S_2O_3$." You will use this glassware to obtain and measure the sodium thiosulfate solution.

3. Obtain about 20 mL of hydrochloric acid solution and about 20 mL of sodium thiosulfate solution from your teacher in the appropriately labeled beakers.

Setting up the CBL System

4. Plug the temperature probe into the Channel 1 input of the CBL. Connect the CBL to the graphing calculator by plugging the black link cable into the base of each unit.

5. Turn on both the CBL and the calculator. Press PRGM on the calculator, and select the PHYSCI program.

6. Go to the MAIN MENU, and select SET UP PROBES. Enter "1" as the number of probes. Select TEMPERATURE from the SELECT PROBE menu. Enter "1" as the channel number.

7. From the MAIN MENU, select COLLECT DATA. Select MONITOR INPUT from the DATA COLLECTION menu. Temperature readings will be displayed on the calculator.

Changing the temperature of the reactants

8. Mix warm tap water and cold tap water together in a water bath until the temperature of the water is about 40°C.

9. Use the graduated cylinder labeled "HCl" to measure 5 mL of the HCl solution. Transfer this solution to one of the test tubes, and place the test tube in the rack.

10. Use the graduated cylinder labeled "$Na_2S_2O_3$" to measure 5 mL of the $Na_2S_2O_3$ solution. Transfer this solution to the other test tube, and place the test tube in the rack.

11. Place the entire rack in the water bath. The solutions in both test tubes should be completely immersed in the water, as shown below. Let the two solutions sit in the water bath for 5 minutes.

12. While you are waiting, use the temperature probe to measure the exact temperature of the water. Record this temperature in your data table. After 5 minutes, the two solutions should be the same temperature as the water.

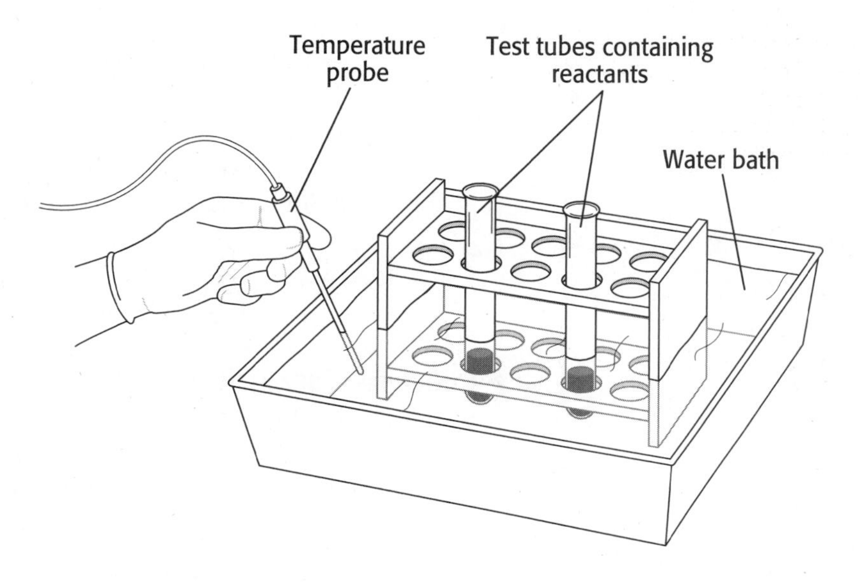

Observing the reaction

13. Carefully pour the contents of one test tube into the other test tube. Start the stopwatch the moment the two reactants come in contact. Carefully pour the contents back and forth between the two test tubes three times.

14. The moment the solution becomes cloudy, stop the stopwatch. Record the elapsed reaction time in your data table. Dispose of the waste as instructed by your teacher.

15. Rinse the test tubes thoroughly with water, and repeat steps 8–14 three more times, adjusting the water to be about 30°C, then about 20°C, and then about 10°C. Be sure to rinse the test tubes thoroughly with deionized water before each reaction.

16. When you have finished, press "+" on the calculator. Then put away all of your materials, clean your work area, and wash your hands thoroughly.

▸ Analyzing Your Results

1. Plot your data on a graph like the one shown on the next page. If you use your graphing calculator, be sure to include a sketch of the graph in your lab report.

2. Describe how the temperature of the reactants affects the time it takes for this reaction to occur.

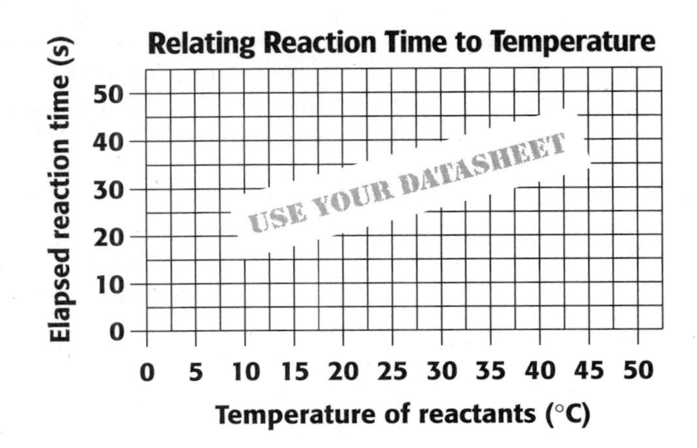

3. Use the graph to estimate the time it would take for this reaction to occur if the reactants were at 5°C. What would the reaction time be if the reactants were at 50°C?

4. Do you think the relationship shown by the graph applies to temperatures below 0°C? (**Hint:** What would happen to the water in the solutions below 0°C?)

▶ Reaching Conclusions

5. Assuming that the reaction of sodium thiosulfate with silver bromide is similar to the reaction you carried out in this experiment, what is the minimum temperature your chemicals must be to ensure that the reaction is complete in 20 s or less?

6. Considering that some of your photo processing chemicals may undergo decomposition reactions when they are stored at room temperature, what is one thing you can do to slow these reactions and extend the shelf life of your chemicals?

▶ Defending Your Conclusions

7. Suppose someone tells you your results are not valid because you did not test the actual reaction used to develop film. What can you do to show that your results are valid?

▶ Expanding Your Knowledge

1. Another factor that affects the rate of a reaction is how dilute a solution of a reactant is. To test the effects of this factor, repeat the same reaction at room temperature, varying the concentration of the HCl solution each time. For the first trial, combine 5 mL of each solution. For the next trial, add 3 mL of deionized water to 2 mL of the HCl solution, and combine this solution with 5 mL of the $Na_2S_2O_3$ solution. For the last trial, add 4 mL of deionized water to 1 mL of the HCl solution, and combine this solution with 5 mL of the $Na_2S_2O_3$ solution. Make a graph of your data, and explain how changing this factor affects the rate of the reaction.

LAB

6 SOLUTIONS

Determining the Concentration of an Ionic Solution

CBL

CHAPTER 6

MATERIALS

- ✓ conductivity probe (and adapter cable)
- ✓ CBL
- ✓ TI graphing calculator and black link cable
- ✓ 100 mL beakers (8)
- ✓ several NaCl solutions of different concentrations
- ✓ 100 mL graduated cylinder
- ✓ deionized water
- ✓ rinse bottle filled with deionized water
- ✓ large beaker
- ✓ tissue paper
- ✓ ring stand
- ✓ small clamp
- ✓ graph paper (optional)

Wear safety goggles, gloves, and a laboratory apron while performing this experiment.

If a chemical gets on your skin or clothing or in your eyes, rinse the affected area immediately with lukewarm water, and alert your teacher.

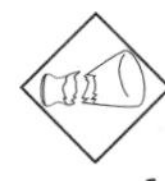

If a piece of glassware breaks, do not pick it up with your bare hands. Place broken glass in a specially designated disposal container.

▸ Objectives

- ***Measure*** the conductivities of several NaCl solutions of known concentrations.
- ***Relate*** your conductivity measurements to the concentrations of the solutions.
- ***Determine*** whether two solutions of NaCl have been labeled with the right concentrations.

▸ Identifying the Problem

Imagine that there has been a mix-up in the lab that prepares saline (sodium chloride, NaCl) solutions for the area hospitals. Lab engineers suspect that two saline solutions may have been labeled with the wrong concentrations. The concentrations of these solutions must be right because serious problems can occur if a saline solution with the wrong concentration is injected into a patient's bloodstream. Imagine that you have been asked to determine the NaCl concentration of each of the saline solutions in question to see if they have been labeled correctly.

▸ Finding Out More Information

When an ionic compound is dissolved in water, ions break away from the solid and spread evenly throughout the solution. This is shown below for sodium chloride, or table salt.

$$\textbf{NaCl (solid)} \rightarrow \textbf{Na}^+ \textbf{ (dissolved)} + \textbf{Cl}^- \textbf{ (dissolved)}$$

Ions dissolved in a solution, like Na^+ and Cl^-, can conduct electricity because they are mobile and carry an electric charge. The conductivity of a solution, or how easily electricity passes through the solution, depends on the concentration of ions.

If a large amount of NaCl is added to water, the solution will have many Na^+ ions and Cl^- ions when the solid dissolves. This solution will conduct electricity well. If less NaCl is added to the same volume of water, the solution will have fewer ions and will therefore be less conductive. Conductivity is measured in SI units of microsiemens per centimeter, or μS/cm. All you need to know is that the greater the conductivity reading is, the more ions there must be in the solution.

▸ Coming Up with a Plan

To determine how conductivity relates to the concentration of NaCl in a solution, you must measure the conductivities of several NaCl solutions of known concentrations. Then make a graph of your data. That way, when you measure the conductivities of the unknown solutions, you can compare their conductivity readings with those that are on the graph. This will allow you to estimate the concentrations of the unknown solutions.

When you are done with the experiment, dispose of the waste as instructed by your teacher, and wash your hands thoroughly.

▸ Performing the Experiment

Preparing for the experiment

1. Prepare tables in your lab report similar to the ones shown below to record your data.

Table 1 Conductivity Data of Standards

Beaker number	Concentration of standard solution	Conductivity (μS/cm)
1	0.50% NaCl	
2	0.40% NaCl	
3	0.30% NaCl	USE YOUR DATASHEET
4	0.20% NaCl	
5	0.10% NaCl	
6	0% NaCl (deionized water)	

Table 2 Conductivity Data of Unknowns

Beaker number	Concentration of solution	Conductivity (μS/cm)
7	Unknown (Labeled as "0.25% NaCl.")	USE YOUR DATASHEET
8	Unknown (Labeled as "0.45% NaCl.")	

2. Obtain eight 100 mL beakers. Label them 1–8.
3. Add 50 mL of each standard solution to beakers 1–6, as indicated in **Table 1.**
4. Pour each of the saline solutions that might have been labeled with the wrong concentrations in beakers 7 and 8. Pour the solution labeled "0.25% NaCl" in beaker 7 and the solution labeled "0.45% NaCl" in beaker 8.

Setting up the CBL System

5. Plug the conductivity probe into the Channel 1 input of the CBL. Connect the CBL to the graphing calculator by plugging the black link cable into the base of each unit. Set the switch on the probe amplifier box to 0–20 000 μS.
6. Turn on both the CBL and the calculator. Press PRGM on the calculator, and select the PHYSCI program.
7. Go to the MAIN MENU, and select SET UP PROBES. Enter "1" as the number of probes. Select CONDUCTIVITY from the SELECT PROBE menu. Enter "1" as the channel number.
8. Select USE STORED from the CALIBRATION menu. Select 0–20 000 MICS (μS) from the CONDUCTIVITY menu.

9. From the MAIN MENU, select COLLECT DATA. Select MONITOR INPUT from the DATA COLLECTION menu.

Measuring the conductivity of the standard solutions

NOTE: *Be sure to take your readings from the calculator and not the CBL.*

10. Starting with beaker 6 (deionized water), submerge the probe in the liquid until the window on the side of the probe is submerged, and gently swirl. Once the reading has stabilized, record the conductivity in **Table 1.**

11. Repeat step 10 for the solutions in beakers 5, 4, 3, 2, and 1, in that order. Record each conductivity reading in **Table 1.** To minimize the chance of contaminating your standards, rinse the probe thoroughly with deionized water over a large beaker, and gently blot the probe dry with tissue paper before testing each solution.

Measuring the conductivity of your unknown samples

12. After rinsing and drying the probe, submerge it in one of your unknown saline solutions and gently swirl. Once the reading has stabilized, record the conductivity in **Table 2.** Do the same for the second unknown solution.

13. When you have finished, dispose of the solutions as instructed by your teacher, and press "+" on the calculator. Then put away all of your materials, clean your work area, and wash your hands thoroughly.

▶ Analyzing Your Results

1. Plot your data from **Table 1** on a graph like the one shown below. If you use your graphing calculator, be sure to include a sketch of the graph in your lab report.

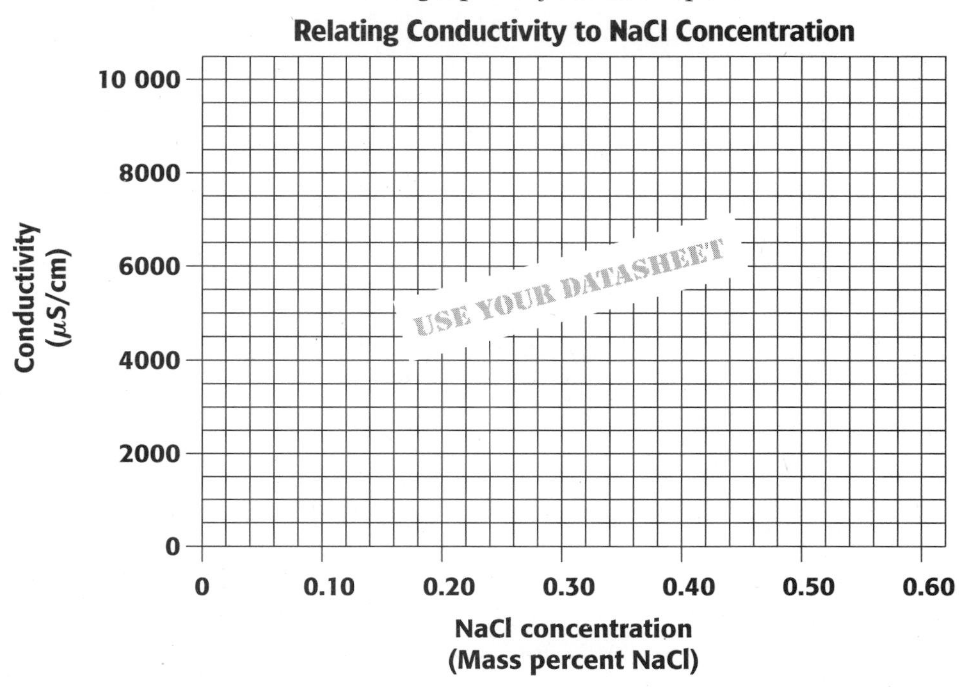

2. Compare the conductivity readings for the unknown solutions in beakers 7 and 8 with the conductivity readings of the standard solutions. Use your graph to determine the NaCl concentrations of the two unknown solutions.

▶ Reaching Conclusions

3. Are the two saline solutions in question labeled correctly?

4. Suppose that one of the unknown solutions had a conductivity reading much higher than that of your highest standard. Would you have been able to use your graph to determine its concentration? Explain why or why not.

▶ Defending Your Conclusions

5. A classmate of yours wants to run a similar experiment to determine the concentration of sugar, a compound made of molecules, in a solution. Will such an experiment work? Explain why or why not.

▶ Expanding Your Knowledge

1. Monitor the conductivity of a local lake, river, or stream for an extended period of time. Make a hypothesis regarding how rainfall or other factors might affect the water's conductivity, and then test your hypothesis.

2. Design an experiment to compare the conductivities of several brands of prepared soups with the mass of sodium listed on the nutritional information label for each soup. Then carry out your experiment.

LAB

7 NUCLEAR CHANGES

Determining the Effective Half-life of Iodine-131 in the Human Body

MATERIALS

- ✓ TI graphing calculator (or computer with graphing software or graph paper and a pencil)

Wear safety goggles while performing this experiment.

If you are working in the laboratory, wear gloves and a laboratory apron.

When you are done with the experiment, wash your hands thoroughly.

▶ Objectives

- ***Make a graph*** from research data that shows the decay activity of the radioactive isotope iodine-131 in the human body.
- ***Analyze*** the graph to find out how iodine-131 is taken up by the thyroid gland.
- ***Calculate*** the effective half-life of iodine-131 in the body.

▶ Identifying the Problem

Doctors sometimes administer a tiny dose of a radioactive isotope of iodine, iodine-131, to study the uptake of iodine by the thyroid gland. When administering iodine-131 to a patient, doctors must take into account the fact that a patient's body may not take up all of the isotope. In addition, not all of the isotope that is taken up will end up in the thyroid gland.

Doctors need to know the *effective half-life* of the isotope in the body to give a proper dose of it to their patients. The effective half-life will be lower than the actual half-life, which is 8.07 days. Imagine that you have been asked by a team of doctors to evaluate their data to determine the effective half-life of iodine-131 in the human body.

▶ Finding Out More Information

Iodine-131 is one of the fission products uranium-235 produces as it decays. Iodine-131 decays by emitting a beta particle, forming a stable isotope of xenon.

For your thyroid gland to produce hormones that your body needs, your thyroid gland must take up iodine. The iodine-131 that doctors administer to patients is usually in the form of sodium iodide, NaI, that is taken orally. The uptake of radioactive iodine is then measured with an instrument that is placed close to the throat, where the thyroid gland is located. This instrument counts the number of beta particles being emitted. The decay rate in counts per minute, or cpm, is related to how much iodine-131 is present.

▶ Coming Up with a Plan

To determine the effective half-life of iodine-131, you will need to evaluate the data from the research study done on 28 individuals. Each individual was given a dose of iodine-131 with an initial decay activity of 18 000 cpm. The averaged data for all 28 individuals appears in the table on the next page. If you make a graph of the data, you can use the graph to predict the effective half-life of iodine-131 in the human body.

CHAPTER 7

Decay Activity for Iodine-131 in the Human Body

Day	Average decay rate (cpm)	Day	Average decay rate (cpm)
0 (start)	154		
1	10 610	11	663
2	8040	12	502
3	6093	13	380
4	4618	14	289
5	3500	15	219
6	2653	16	166
7	2010	17	127
8	1523	18	95
9	1154	19	72
10	875	20	55

▶ Performing the Experiment

Preparing for the experiment

1. Prepare a table in your lab report similar to the one shown below to record your data.

Calculating the Effective Half-life of Iodine-131

Decay rate (cpm)	Time (days)	Elapsed time to halve the decay rate (days)
10 000		
5000		
2500	USE YOUR DATASHEET	
1250		
Average effective half-life		

Setting up the graphing calculator

2. Turn on the graphing calculator. Press STAT. Select CLRLIST from the EDIT menu. Press 2nd "L1," then ",". Then press 2nd "L2," ",", 2nd "L3," and ENTER. This will clear lists L1–L3.

3. Press 2nd MEM. Select DELETE from the MEMORY menu. Select Y-Vars, then press ENTER to delete each Y-Var that is stored in the calculator's memory.

4. Press STAT. Select EDIT from the EDIT menu. The screen will display an empty table with three columns.

Entering the data

5. In the first column, enter the numbers 0–20 to represent the days of the research. Do this by pressing 0, then ENTER, then 1, then ENTER, etc. until you have entered each day.

6. Use the right arrow key to move the cursor to the second column. Enter the average decay rate for each day. For example, for day zero (start of the research), press 154, then ENTER, 10 610, then ENTER, etc. until you have entered each value.

Viewing the graph

7. Press 2nd STAT PLOT. Select PLOTSOFF, and then press ENTER. Press 2nd STAT PLOT again, then select PLOT1. Turn PLOT1 on by selecting ON and pressing ENTER. Use the arrow key to move the cursor to TYPE, and select the line graph (the second option). Press ENTER.
8. Enter "L1" for the Xlist and "L2" for the Ylist. Select the small square for the MARK.
9. Press ZOOM. Use the arrow key to scroll down to ZOOMSTAT, and press ENTER. You will see a graph of the data.
10. Press TRACE, and use the arrow keys to move across the graph. As you do, the x- and y-values will be displayed at the bottom of the screen.
11. Press "Y=." For Y_1, enter 10 000. Use the down arrow key to move to Y_2, and enter 5000. Move down to Y_3, and enter 2500. Then move down to Y_4, and enter 1250.
12. Press GRAPH. The graph should now have four horizontal lines across it. Each line is half as high as the one above it. You will use these lines to estimate the effective half-life of iodine-131 in the human body.

▶ Analyzing Your Results

1. Use your graph to describe how the concentration of iodine in the thyroid glands of the patients changed over the 20-day research period.
2. When did the thyroid glands of the patients have the maximum concentration of iodine-131?
3. Press TRACE. Use the arrow keys to estimate (to the nearest 0.1 day) the time at which the average decay rate was 10 000 cpm. Use the horizontal line at 10 000 cpm to help you estimate. Enter your answer in your data table. Do the same to estimate the time at which the average decay rate was 5000 cpm, 2500 cpm, and 1250 cpm. Record your answers in your data table.
4. Calculate how long it took (to the nearest 0.1 day) for the average decay rate to drop from 10 000 cpm to 5000 cpm. Do this by subtracting the time at which the decay rate was 10 000 cpm from the time at which the decay rate was 5000 cpm. Record this value in your data table.

5. Repeat item 4 to find out how many days it took for the average decay rate to drop from 5000 cpm to 2500 cpm and then from 2500 cpm to 1250 cpm. Record each of your answers in your data table. These values and that from item 4 represent effective half-lives for iodine-131 in the body.

6. Calculate the average effective half-life of iodine-131 by adding the three effective half-lives you just found, and then dividing by 3. Record your answer in your data table.

▶ Reaching Conclusions

7. Explain why the concentration of iodine-131 was low initially, increased quickly, and then slowly decreased over time.

8. The actual half-life of iodine-131 is 8.07 days. How does this value compare with the average effective half-life that you calculated? Why is the effective half-life different from the actual half-life?

9. If the liver absorbs one-tenth as much iodine as the thyroid gland, how would the graph have looked if the study had been done on the liver instead of the thyroid gland?

▶ Defending Your Conclusions

10. In this study, the individuals tested each had similar results. Suppose the data had shown that the amount of iodine-131 in the thyroid varied considerably from person to person. Would these results change the way doctors use the data to determine a specific dose? Explain.

▶ Expanding Your Knowledge

1. Research the effects of nuclear weapons testing and the accident at Chernobyl that occurred in the Ukraine in 1986. Make a poster outlining how iodine-131 makes its way through the environment, eventually reaching humans.

2. Research how spent fuel rods from nuclear power plants are disposed of. In groups of four, develop a proposal for the safe storage of waste generated from nuclear power plants. Prepare a panel discussion to present and defend your proposal to your classmates.

LAB

8 FORCES

Determining Your Acceleration on a Bicycle

MATERIALS

- ✓ force sensor (and adapter cable)
- ✓ CBL
- ✓ TI graphing calculator and black link cable
- ✓ multi-gear bicycle
- ✓ table support
- ✓ object with a mass of 2–6 kg
- ✓ string
- ✓ sturdy table
- ✓ short piece of rope
- ✓ bathroom scale
- ✓ graph paper (optional)
- ✓ calculator (optional)

Wear safety goggles and gloves while performing this experiment.

If you are working in the laboratory, wear a laboratory apron.

Make sure that the bicycle is secured to a lab table at all times. Do not lean the bicycle against anything unless it is held in place with rope. The bicycle should be supported by someone at all times.

Be sure to keep your fingers, hair, clothing, and jewelry away from the spokes and gear-wheels as you change gears.

When you are done with the experiment, wash your hands thoroughly.

▶ Objectives

- ***Design*** a system that can simulate the forces acting on a bicycle as it is being ridden.
- ***Measure*** the forward, or impelling, force acting on the bicycle for each gear.
- ***Calculate*** the acceleration that is possible for each gear by using Newton's second law of motion.
- ***Determine*** which gears provide the greatest acceleration.

▶ Identifying the Problem

Imagine that you are training for an upcoming bicycle race. To be competitive in the race, you will have to use lower gears when accelerating or traveling uphill and higher gears when maintaining a certain speed or traveling downhill.

But because the gear ranges sometimes overlap, it can be difficult to know when you should change gears. You have decided to test how much acceleration each gear can provide so that you can come up with some efficient shifting strategies for the race.

▶ Finding Out More Information

Newton's second law of motion states that the unbalanced force acting on an object equals the object's mass times its acceleration. This law is written as follows.

$$\textbf{force} = \textbf{mass} \times \textbf{acceleration}$$

If the object's mass is measured in kilograms and its acceleration is in meters per second squared, the units for force are $kg \cdot m/s^2$, or newtons (N). If you know an object's mass and the force that is acting on the object, you can calculate the object's acceleration by using the following equation.

$$\textbf{acceleration} = \frac{\textbf{force}}{\textbf{mass}}$$

▶ Coming Up with a Plan

You might think it would be easy to tell which gears provide the greatest acceleration by riding your bicycle around outside. But because the terrain varies so much, it would be hard to know if you tested every gear under exactly the same conditions. Instead, you should measure the force that acts on the bicycle for each gear in a controlled setting. That way, the conditions remain exactly the same throughout the experiment so that you can compare the gears.

You can hang a small object from one of the pedals to simulate the downward force of your foot as you ride the bicycle. Then you can measure the impelling force, or the force moving the bicycle forward, for each gear. You can then use Newton's second law of motion to determine which gears provide the greatest acceleration.

▶ Performing the Experiment

Preparing for the experiment

1. Prepare tables in your lab report similar to the ones shown below to record your data.

Table 1 Forces and Masses of Objects

Object	Weight (lb)	Force (N)	Mass (kg)	
Bicycle and you				
You				Ratio of your force to hanging object's force
Hanging object				

Note: Your version of **Table 2** might vary slightly from the one shown below, depending on how many gears the bicycle you are testing has.

Table 2 Force and Acceleration Measurements and Calculations

Front gear-wheel	Rear gear-wheel	Impelling force hanging object produces (N)	Your impelling force (N)	Your acceleration on the bicycle (m/s^2)
Smallest	S1 (low)			
	S2			
	S3			
	S4			
	S5			
Middle	M1 (low)			
	M2			
	M3			
	M4			
	M5			
Largest	L1 (low)			
	L2			
	L3			
	L4			
	L5			

2. Use a bathroom scale to weigh yourself in pounds while you are holding the bicycle. Then weigh yourself without holding the bicycle. Record both weights in **Table 1.**

Setting up the CBL System

3. Plug the force sensor into the Channel 1 input of the CBL. Connect the CBL to the graphing calculator by plugging the black link cable into the base of each unit.
4. Turn on both the CBL and the calculator. Press PRGM on the calculator, and select the PHYSCI program.
5. Go to the MAIN MENU, and select SET UP PROBES. Enter "1" as the number of probes. Select FORCE from the SELECT PROBE menu. Enter "1" as the channel number.
6. Select USE STORED from the CALIBRATION menu. Then select STUDENT FORCE from the FORCE PROBE menu.
7. From the MAIN MENU, select COLLECT DATA. Select MONITOR INPUT from the DATA COLLECTION menu. Force readings will be displayed on the calculator.

Setting up the bicycle

8. Use a short piece of rope to connect the bicycle seat post to the force sensor, as shown below. The rope should be level to ensure that no unnecessary force is being placed on the sensor.

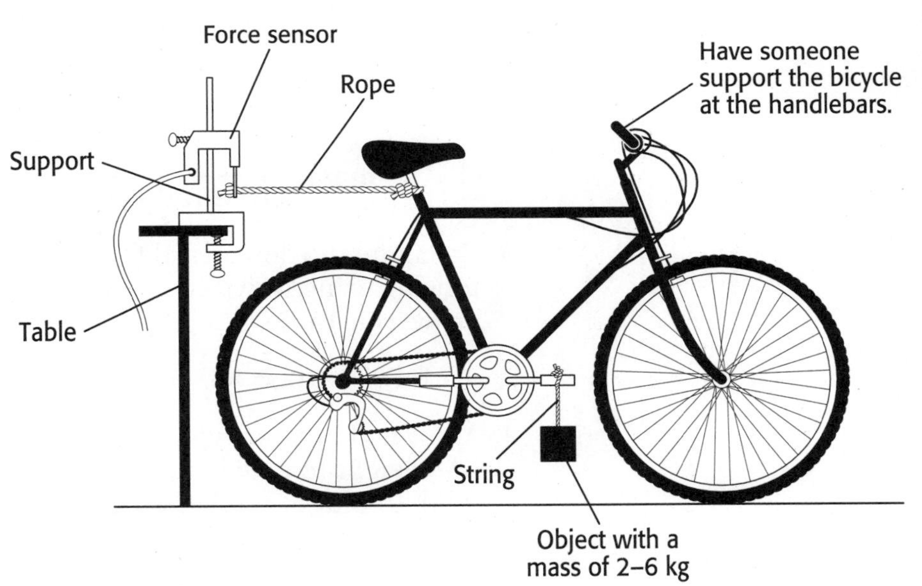

9. Measure the mass (in kg) of the object to be hung from the pedal, and record it in **Table 1.** Use string to hang the object.
10. Put the bicycle in the lowest gear. As you switch gears, one person should lift the bicycle by the seat post so that the rear wheel is off the floor, while a second person turns the pedals and switches gears. A third person should be supporting the bicycle by the handlebars at all times.

11. Put the bicycle's rear wheel back on the floor in its original position, and position the pedals as shown.

Measuring the impelling force the hanging object produces

12. Measure the impelling force, and record this value in **Table 2.** Be sure that the person supporting the handlebars is not putting any horizontal forces on the bicycle.
13. While keeping the chain on the smallest gear-wheel in front, repeat steps 10–12 for each rear gear-wheel.
14. Switch to the next gear in front, and repeat steps 10–12 for each rear gear-wheel. If you have a third gear in front, collect data for this gear also.
15. When you have finished, press "+" on the calculator. Then put away all of your materials, clean your work area, and wash your hands thoroughly.

▶ Analyzing Your Results

1. Convert both your weight and the combined weight of the bicycle and you to newtons by using the following equation. Record your answers in **Table 1.**

$$\textbf{force (N)} = \textbf{weight } (\cancel{\textbf{lb}}) \times \frac{\textbf{1 N}}{\textbf{0.225 } \cancel{\textbf{lb}}}$$

2. Calculate the combined mass of the bicycle and you in kilograms by using the following equation. Record your answer in **Table 1.**

$$\textbf{mass} = \frac{\textbf{force}}{\textbf{free-fall acceleration (9.8 m/s}^2\textbf{)}}$$

3. Calculate the downward force of the hanging object in newtons by using Newton's second law of motion. Record your answer in **Table 1.**

$$\textbf{force} = \textbf{mass} \times \textbf{free-fall acceleration (9.8 m/s}^2\textbf{)}$$

4. Assume that when you ride a bicycle you can use your entire weight to pedal. Calculate the ratio of your downward force to that of the hanging object by using the following equation. Record your answer in **Table 1.**

$$\textbf{ratio of your force to hanging object's force} = \frac{\textbf{your force}}{\textbf{force of hanging object}}$$

5. For each gear, calculate your impelling force by using the following equation. Record your answers in **Table 2.**

$$\textbf{your impelling force} = \textbf{object's impelling force} \times \textbf{ratio of forces}$$

6. For each gear, calculate your acceleration on the bicycle by using the following equation. Record your answers in **Table 2.**

$$\textbf{acceleration} = \frac{\textbf{your impelling force}}{\textbf{combined mass of bicycle and you}}$$

7. Plot your acceleration data from **Table 2** on a bar graph like the one shown below.

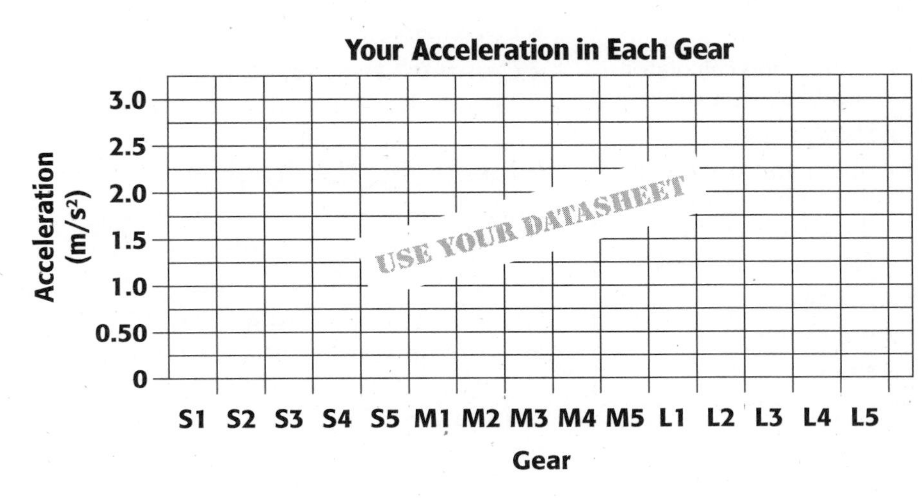

▶ Reaching Conclusions

8. Use your bar graph to list the gears in order from the one that provides the greatest acceleration to the one that provides the least acceleration.

9. Which gear(s) on the smallest front gear-wheel overlap with those on the middle front gear-wheel? Which gear(s) on the middle front gear-wheel overlap with those on the largest front gear-wheel?

▶ Defending Your Conclusions

10. In this experiment, you calculated the force you would exert on the bicycle by assuming that you could put all of your weight on the pedals. Is this a realistic assumption? How might your results have been different if you were able to measure the actual force you could exert while riding the bicycle?

▶ Expanding Your Knowledge

1. One of the simplest ways bicyclists compare gears is by using the following equation.

$$\textbf{distance} = \textbf{diameter of rear wheel} \times \pi \times \frac{\textbf{number of teeth in front gear-wheel}}{\textbf{number of teeth in rear gear-wheel}}$$

This equation calculates the distance you can travel for each revolution of the pedals. Use this equation to calculate the distance in meters you could travel in each gear for a single turn of the pedals.

LAB

9 WORK

Determining Which Ramp Is More Efficient

MATERIALS

- ✓ force sensor (and adapter cable)
- ✓ CBL
- ✓ TI graphing calculator and black link cable
- ✓ sturdy chair
- ✓ board approximately three times as long as the chair's height
- ✓ several heavy books
- ✓ rolling cart
- ✓ meterstick
- ✓ string

Wear safety goggles and gloves while performing this experiment.

If you are working in the laboratory, wear a laboratory apron.

Place the rolling cart on a table when it is not being used. Don't leave it on the floor where someone might step on it.

When you are done with the experiment, wash your hands thoroughly.

▶ Objectives

- ***Measure*** the force required to pull an object up a ramp under different conditions.
- ***Calculate*** work input and work output.
- ***Compare*** the mechanical advantage and efficiency of two ramps that have different lengths.

▶ Identifying the Problem

Speedy Shipping Company has just purchased 10 new trucks for transporting large appliances. Because the drivers will have to move the appliances in and out of the trucks by themselves, a supervisor is planning to equip each truck with a ramp to make loading and unloading easier.

If the ramps are too short, they will not help the drivers move appliances up the ramp. But if the ramps are too long, they will no longer be practical. Imagine that you have been asked to compare the mechanical advantage and efficiency of two different ramps that are being considered by the supervisor.

▶ Finding Out More Information

The work needed to lift an object is equal to the object's weight, or force, times the distance it is lifted, as shown below.

$$\textbf{work} = \textbf{force} \times \textbf{distance}$$

If force is measured in newtons (N) and distance is measured in meters (m), work is expressed in units of N•m, or joules (J).

When heavy objects need to be lifted, an inclined plane, or ramp, can make the job easier. That's because the same amount of work can be spread out over a longer distance. Less force will be required overall. Mechanical advantage is a measure of how much easier a machine makes work.

Efficiency is a measure of how the work put in to a process compares with the work done by the process. The efficiency of any simple machine is the ratio of useful work output to total work input. No simple machine, including a ramp, can be 100 percent efficient because some work input is always lost to friction.

▶ Coming Up with a Plan

To compare two ramps, you need to determine how the work input differs using each ramp. In this lab, you will do work on a rolling cart. Work input equals the force needed to pull the cart up the ramp (which you can measure using the force sensor) times the length of the ramp. Work output is the same for both ramps and is equal to the height the cart is lifted times the cart's weight.

You can also investigate how friction affects the force needed to pull the cart up each ramp. To maximize the effects of friction, you can turn the cart upside down and slide it up the ramp without using its wheels. To minimize the effects of friction, you can roll the cart up the ramp using its wheels.

▶ Performing the Experiment

Preparing for the experiment

1. Prepare tables in your lab report similar to the ones shown below to record your data.

Table 1 Work Output for Both Ramps

Height of chair (to seat) (m)	Weight of cart (N)	Work output (J)

Table 2 Work Input, Mechanical Advantage, and Efficiency of Each Ramp

Ramp length (m)	Sliding or rolling the cart?	Force (N)	Work input (J)	Mechanical advantage	Efficiency (%)
	Sliding				
	Rolling				
	Sliding				
	Rolling				

Setting up the CBL System

2. Plug the force sensor into the Channel 1 input of the CBL. Connect the CBL to the graphing calculator by plugging the black link cable into the base of each unit.
3. Turn on both the CBL and the calculator. Press PRGM on the calculator, and select the PHYSCI program.
4. Go to the MAIN MENU, and select SET UP PROBES. Enter "1" as the number of probes. Select FORCE from the SELECT PROBE menu, and enter "1" as the channel number.
5. Select USE STORED from the CALIBRATION menu. Then select STUDENT FORCE from the FORCE PROBE menu.
6. From the MAIN MENU, select COLLECT DATA. Select MONITOR INPUT from the DATA COLLECTION menu. Force readings will be displayed on the calculator.

Setting up the longer ramp

7. Place the back of the chair against a wall, and position the board so that it forms a ramp from the floor to the seat of the chair, as shown below. Extend the ramp as far out as possible. Place several heavy books at the base of the ramp to keep it from sliding.

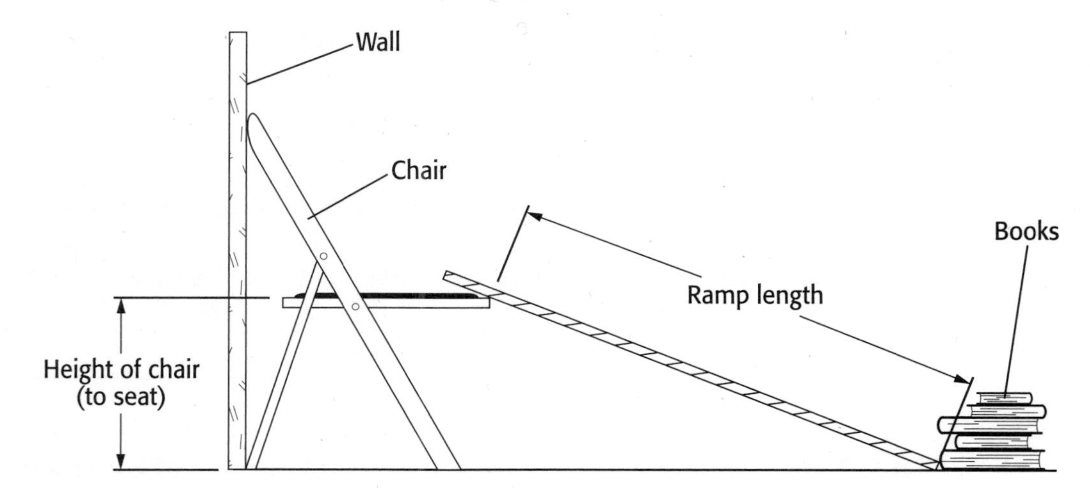

Determining work output

8. Use the force sensor to measure the weight of the cart to the nearest 0.1 N by hanging the cart directly from the force sensor. Once the reading has stabilized, record the force in **Table 1.**

9. Use a meterstick to measure the height of the chair (from the floor to the seat of the chair) in meters, as shown. Record this height in **Table 1.**

Determining work input using the longer ramp

10. Measure the length of the ramp in meters. Measure from where the ramp meets the floor to where the ramp meets the seat of the chair, as shown. Record this length in **Table 2.**

11. Attach the force sensor to the cart with a piece of string. Turn the cart upside down so that its wheels are up. Holding on to the force sensor, slide the upside down cart slowly and steadily up the ramp. Practice this motion several times.

12. When the cart is halfway up the ramp on one of your practice runs, have a lab partner record the force reading to the nearest 0.5 N.

13. Repeat step 12 once more. If this second force reading is close to the first one, then your measurements are precise. Record this second force reading in **Table 2.** If this reading is not close to the first one, you need to practice being consistent about how you pull the cart up the ramp.

14. Turn the cart over so that it can roll on its wheels. Pull the cart up the ramp again, this time using its wheels. Measure the force required, and record it in **Table 2.**

Determining work input using the shorter ramp

15. Now make a shorter, steeper ramp by adjusting the base of the board to be about twice as close to the chair as it was before. Secure the base of the ramp with several heavy books.
16. Measure the length of this shorter ramp in meters. Measure from where the ramp meets the floor to where the ramp meets the seat of the chair. Record this length in **Table 2.**
17. First measure the force required to slide the cart steadily up the ramp without using its wheels. Then turn the cart over and measure the force required to roll it up the ramp using its wheels. Record both force readings in **Table 2.**
18. When you have finished, press "+" on the calculator. Then put away all of your materials, clean your work area, and wash your hands thoroughly.

▶ Analyzing Your Results

1. Calculate the work output for both ramps in joules by using the following equation. Record your answer in **Table 1.**

   $$\text{work output} = \text{height of chair (to seat)} \times \text{weight of cart}$$

   This is the work required to lift the cart from the floor to the seat of the chair without using any machines and without any friction.

2. Use the following equation to calculate the work input required using each ramp, first when sliding and then when rolling the cart. Record your answers in **Table 2.**

   $$\text{work input} = \text{force} \times \text{ramp length}$$

   This is the work required to move the cart from the floor to the seat of the chair using the ramp.

3. Calculate the mechanical advantage of each ramp by using the following equation. Record your answers in **Table 2.**

   $$\text{mechanical advantage} = \frac{\text{ramp length}}{\text{height of chair (to seat)}}$$

4. Use the following equation to calculate the percent efficiency of each ramp, first when sliding and then when rolling the cart. Record your answers in **Table 2.**

   $$\text{percent efficiency} = \frac{\text{work output}}{\text{work input}} \times 100$$

CHAPTER 9

▶ Reaching Conclusions

5. Which ramp has a greater mechanical advantage? How does the length of a ramp affect its mechanical advantage? For the Speedy Shipping Company, would you recommend a ramp with a greater or lesser mechanical advantage?
6. Which ramp is more efficient? In your own words, explain to the supervisor why the ramp is more efficient.
7. In each case, the work output is less than the work input. Explain why.
8. Suppose that a driver needs to load an appliance weighing 1010 N into the back of a truck 1.0 m off the ground. Speedy Shipping Company does not allow its employees to exert more than 250 N of force. How long of a ramp should the driver use? Assume that the driver can use a rolling cart to reduce friction.

▶ Defending Your Conclusions

9. Suppose someone tells you that your results are not valid because you measured the force needed to pull a light rolling cart up the ramp instead of a heavy appliance. Are your results still valid? Explain your answer.

▶ Expanding Your Knowledge

1. Research the efficiencies of several brands of refrigerators or other household appliances. Summarize your results in a report.

LAB

10 HEAT AND TEMPERATURE

Determining the Better Insulator for Your Feet

MATERIALS

- ✓ 2 temperature probes
- ✓ CBL
- ✓ TI graphing calculator and black link cable
- ✓ 2 plastic beverage bottles
- ✓ 2 rubber stoppers that fit temperature probes and bottle openings
- ✓ cotton sock
- ✓ wool sock
- ✓ warm tap water at 40–45°C
- ✓ balance
- ✓ cold tap water
- ✓ paper towels
- ✓ stopwatch

Wear safety goggles and gloves while performing this experiment.

If you are working in the laboratory, wear a laboratory apron.

When you are done with the experiment, wash your hands thoroughly.

▶ Objectives

- ***Model*** a hiker's feet in wet conditions using bottles of water covered with wet socks.
- ***Measure*** the temperature changes.
- ***Calculate*** the energy transferred away from each "foot" as heat in joules by using the specific heat equation.
- ***Compare*** the effectiveness of the two socks as insulators in wet conditions.
- ***Evaluate*** which material insulates better.

▶ Identifying the Problem

Imagine that the editor of an outdoor magazine has hired you to be a technical consultant for an article on choosing the best socks to wear when hiking. The editor is particularly interested in how well two different socks insulate the feet in wet conditions. She has given you one sock made of cotton and another sock made of wool. You have been asked to test the insulating abilities of both socks when they are wet and then determine which material offers better insulation. You must also supply any data that supports your findings so that it can be included in the article.

▶ Finding Out More Information

Energy that is transferred as heat is always transferred from a warmer object to a cooler object. People wear insulators, like socks, gloves, hats, and other clothing, when it is cold outside to reduce the amount of energy transferred away from their warm bodies as heat.

If a person is wearing socks and other articles of clothing that are good insulators, the person feels warmer because he or she is able to slow down the transfer of energy away from the body as heat. However, if the person goes outside with little insulation, he or she feels colder because energy is transferred away from the body as heat much faster.

▶ Coming Up with a Plan

Your first thought about how to do this experiment might be to use your own foot as a model. But even if you were able to "feel" which sock made your foot warmer, you wouldn't be able to measure exactly how much better it was at insulating your foot. You also wouldn't have the data you need to support your findings.

Instead, you need to make measurements. You can't measure energy directly, but you can measure temperature over a period of time to see how it changes for each foot. Then you can use the specific heat equation to calculate the energy transfer as heat that must have occurred.

You can use a bottle of warm water as a model of a foot because a real foot is also made mostly of water. Measuring the temperature of the water inside a bottle is also much easier than measuring the temperature inside your foot.

▶ Performing the Experiment

Preparing for the experiment

1. Prepare tables in your lab report like those shown below.

Table 1 Mass of Warm Water Used to Make "Feet"

Bottle number	Mass of bottle (g)	Mass of bottle and water (g)	Mass of water (g)	Mass of water (kg)
1		USE YOUR DATASHEET		
2				

Table 2 Temperature Changes of Each "Foot"

Bottle number/sock material	Initial temp. (°C)	After 4 min (°C)	After 8 min (°C)	After 12 min (°C)	Final temp. (°C)	Δt (°C)	Energy transferred as heat (J)
1/cotton			USE YOUR DATASHEET				
2/wool							

2. Obtain two empty plastic bottles. Label one of the bottles 1 and the other bottle 2.

Setting up the CBL System

3. Plug one of the temperature probes into the Channel 1 input of the CBL. Plug the other temperature probe into the Channel 2 input of the CBL. Connect the CBL to the graphing calculator by plugging the black link cable into the base of each unit.
4. Turn on both the CBL and the calculator. Press PRGM on the calculator, and select the PHYSCI program.
5. Go to the MAIN MENU, and select SET UP PROBES. Enter "2" as the number of probes. Select TEMPERATURE from the SELECT PROBE menu. Enter "1" as the channel number. Select TEMPERATURE again from the SELECT PROBE menu, and enter "2" as the channel number.
6. From the MAIN MENU, select COLLECT DATA. Select MONITOR INPUT from the DATA COLLECTION menu. The temperature readings for both probes will be displayed on the calculator.

Making a model of a hiker's feet

7. Measure the mass of each bottle in grams. Record each mass in **Table 1.**

8. Fill each bottle nearly to the top with warm tap water. Keep in mind that when the temperature probe and rubber stopper are inserted, no water should run out. In addition, there should be very little air space left at the top of the bottle. If you need to, experiment to determine the proper level of water. Then fill each bottle to this level with warm water.

9. If there is water on the outside of the bottles, use paper towels to dry them. Measure the mass of each bottle filled with warm tap water. Record each mass in **Table 1.**

10. Carefully insert one of the temperature probes into one of the rubber stoppers. Then insert this stopper into the opening of one of the bottles so that the tip of the temperature probe is at the center of the bottle. Set up the other bottle the same way. If there is water on the outside of the bottles, use paper towels to dry them.

11. Allow each temperature probe 1 minute to adjust to the warm water. Then record the initial temperature of the water in each bottle in **Table 2.**

Comparing the socks in wet conditions

12. Soak both socks in cold tap water. Wring the socks out so that they are thoroughly wet but not dripping. Cover bottle 1 with the cotton sock and bottle 2 with the wool sock.

13. Wait exactly 4 minutes, and then measure the temperature of the water in each bottle again. Record these temperatures in **Table 2.** Continue to record the temperature of the water in each bottle every 4 minutes for a total of 16 minutes.

14. When you have finished, press "+" on the calculator. Then put away all of your materials, clean your work area, and wash your hands thoroughly.

▶ Analyzing Your Results

1. Calculate the mass of water in each bottle in grams by using the following equation. Record your answers in **Table 1.**

$$\text{mass of water} = \text{mass of bottle and water} - \text{mass of bottle}$$

2. Convert each mass of water to kilograms by using the following equation. Record your answers in **Table 1.**

$$\text{mass of water (kg)} = \text{mass of water (g)} \times \frac{1\ \text{kg}}{1000\ \text{g}}$$

3. Calculate the change in temperature, Δt, for the water in each bottle by using the following equation. Record your answers in **Table 2.**

$$\Delta t = \text{final temperature} - \text{initial temperature}$$

4. Calculate the energy transferred away from each "foot" as heat in joules by using the following equation. Use the value 4186 J/kg•K for the specific heat of water. Record your answers in **Table 2.**

$$\textbf{energy} = \textbf{specific heat} \times \textbf{mass of water (kg)} \times \Delta t$$

▶ Reaching Conclusions

5. Which material, cotton or wool, allowed more energy to be transferred away from the "foot" as heat?

6. How significant a difference is there in the insulating abilities of the two wet socks?

7. Which material do you recommend that hikers wear for socks when hiking in the cold rain?

▶ Defending Your Conclusions

8. In this experiment, you used a bottle of warm water to model energy transferred away from a hiker's foot as heat. How is this model different from the energy transferred away from a real foot as heat? How would you defend your results if someone claimed these differences made your results invalid?

▶ Expanding Your Knowledge

1. Using a similar model, design and perform an experiment that compares the insulating abilities of several different building materials.

2. Consult your local building materials store or library, or use the Internet to research the latest insulation developments in the construction or clothing industry. What kinds of materials are currently being developed and used?

3. Newer synthetic polyester clothing is advertised as having good insulating abilities in wet conditions. Test one or more of these synthetic materials against wool or cotton using your model.

4. Research the structure of wool to find out why it works so well in wet conditions.

LAB 11 WAVES

CHAPTER 11

Tuning a Musical Instrument

MATERIALS

- ✓ microphone (CBL-compatible)
- ✓ CBL
- ✓ TI graphing calculator and black link cable
- ✓ TI-Graph Link cable (optional)
- ✓ computer with a printer (optional)
- ✓ calculator (optional)
- ✓ any stringed musical instrument (guitar, mandolin, banjo, violin, etc.)

Wear safety goggles while performing this experiment.

If you are working in the laboratory, wear gloves and a laboratory apron.

Treat the musical instrument with care. Be careful not to use excessive force when plucking the strings.

When you are done with the experiment, wash your hands thoroughly.

▶ Objectives

- ***Relate*** the pitch of a musical note to its frequency.
- ***Apply*** the concept of wave interference to tune an instrument using an electronic device.
- ***Compare*** your ability to tune an instrument using an electronic device with your ability to tune an instrument by ear.

▶ Identifying the Problem

Musicians often debate the value of using electronic devices to tune their instruments. Some musicians find electronic tuning devices very useful, especially in noisy environments. Others believe that the human ear does a better job, so they tune their instruments using their ability to distinguish between pitches. Imagine that you have been asked to try tuning an instrument both ways to find out which way you prefer.

▶ Finding Out More Information

If two sounds are in tune, the sound waves are identical and interfere constructively. The sound you hear is the sum of the two sound waves. An electric signal made from the resulting sound wave looks like the graph in **Figure 11-1.**

Figure 11-1

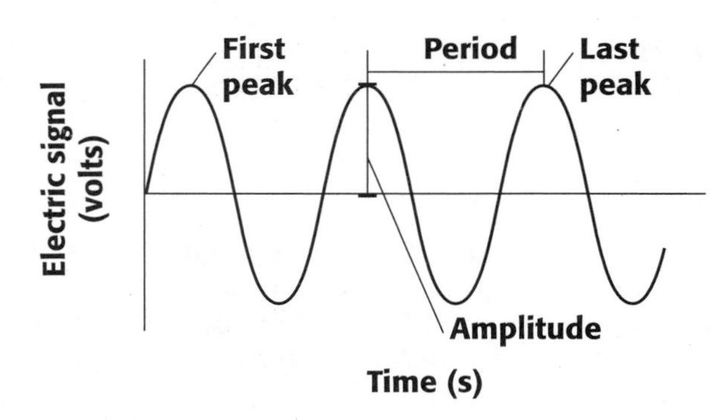

If two sounds are not in tune, the sound waves interfere both constructively and destructively. The amplitude (volume) of the resulting sound wave varies, as shown in **Figure 11-2.** What you hear is a series of alternating loud and soft sounds called beats.

Figure 11-2

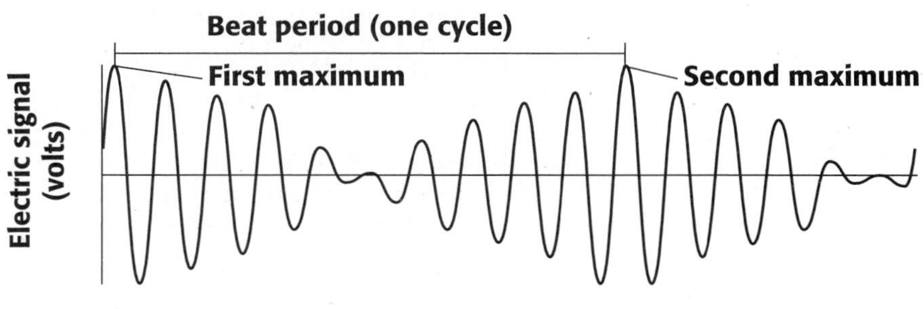

▶ Coming Up with a Plan

You can determine if the pitch of a sound is high or low just by listening. But to determine the number of vibrations that occur over time— the frequency of the sound wave—you need to see a graph of the sound wave. If you use a microphone, you can convert a sound into an electric signal. A graphing calculator can then generate a graph showing how the electric signal changes over a period of time.

The microphone and graphing calculator are the electronic devices you will use to tune your instrument. If your instrument is out of tune, the graph will look like the graph shown in **Figure 11-2.** You must adjust the instrument until the graph looks similar to the one shown in **Figure 11-1.** Then you can tune your instrument again using only your ability to distinguish between pitches.

▶ Performing the Experiment

Preparing for the experiment

1. Prepare tables in your lab report similar to the ones shown below to record your data.

Table 1 Two Different Sounds Played Separately (Sounds 1 and 2)

Sound	Amplitude (V)	Time of first peak (s)	Time of last peak (s)	Elapsed time (s)	Number of waves between first and last peaks	Period (s)	Frequency (Hz)
1			USE YOUR DATASHEET				
2							

Table 2 Two Sounds Out of Tune that Are Played Together (Sound 3)

Maximum amplitude (V)	Time of first maximum (s)	Time of last maximum (s)	Elapsed time (s)	Number of cycles between first and last maximums	Beat period (s)	Frequency (Hz)
		USE YOUR DATASHEET				

Setting up the CBL System

2. Plug the microphone into the Channel 1 input of the CBL. Connect the CBL to the graphing calculator by plugging the black link cable into the base of each unit.

3. Turn on both the CBL and the calculator. Press PRGM on the calculator, and select the PHYSCI program.

4. Go to the MAIN MENU, and select SET UP PROBES. Enter "1" as the number of probes. Select MICROPHONE from the SELECT PROBE menu. Enter "1" as the channel number.

Comparing two different sound waves

5. With the help of your lab partners, hold the microphone close to your instrument and play a single note. Quickly press ENTER on the calculator to record this note. To generate a smooth graph, you may have to experiment by placing the microphone in different places.
6. Use the arrow keys to trace the points on the graph. The x-value (time) and y-value (electric signal) for each point will be displayed on the bottom of the screen. Determine the time in seconds of the first peak (not counting the very first tall peak) and that of the last peak shown on the graph. Record these times in **Table 1.**
7. Count the number of waves between the first peak and the last peak. Record this number in **Table 1.**
8. Trace the points on the graph to determine the maximum y-value, or the amplitude, of the wave in volts. Record the amplitude in **Table 1.**
9. Print a copy of this wave using the TI-Graph Link cable, or make a sketch of it in your lab report. Label it "Sound 1."
10. Press ENTER, and select YES to repeat. Hold the microphone close to your instrument, and play a note that is noticeably higher in pitch than the previous note. Quickly press ENTER, and generate a graph of the wave as before.
11. Repeat steps 6–9, labeling this graph "Sound 2."
12. Press ENTER, and select NO when asked if you want to repeat. Return to the MAIN MENU.

Producing a beat

13. From the MAIN MENU, select SET UP PROBES. Select VOLTAGE from the SELECT PROBE menu. Enter "1" as the channel number. Select USE STORED from the CALIBRATION menu.
14. From the MAIN MENU, select COLLECT DATA. Select TIME GRAPH from the DATA COLLECTION menu. Enter "0.01" as the time between samples in seconds. Enter "99" as the number of samples. Press ENTER, and select USE TIME SETUP.
15. Hold the microphone close to your instrument, and play two closely related sounds. Quickly press ENTER. It works best if you choose a pitch on one string that closely matches the pitch on the next open string (for example, the fifth fret on the E string [the same as an A] of a guitar matches the open note on the A string). The graph generated should look similar to the one shown in **Figure 11-2.**

16. Use the arrow keys to trace the points on the graph. Determine the time of the first maximum and that of the last maximum. Record these times in **Table 2.**

17. Count the number of cycles (refer to **Figure 11-2**) between the first maximum and the last maximum. Record this number in **Table 2.**

18. Trace the points on the graph to determine the maximum *y*-value, or the amplitude, of one of the maximums. Record the amplitude in **Table 2.**

19. Print a copy of this wave using the TI-Graph Link cable, or make a sketch of it in your lab report. Label it "Sound 3."

Tuning your instrument first electronically, then by ear

20. Press ENTER, and select YES to repeat. Use the same two notes you used in the previous section. Have your lab partner put the open string slightly out of tune.

21. With the microphone close, simultaneously strike the fretted note (fifth fret on the E string) and the note on the open string (A string). Quickly press ENTER, and look for beats.

22. Press ENTER, and select YES to repeat. Raise the pitch of the open string slightly. Repeat step 21. If the beat maximums are closer together, then you changed the pitch in the wrong direction. If the beat maximums are farther apart but still visible, then you are closer to the correct pitch.

23. Continue to adjust the open string and repeat step 21 until beats are no longer visible. Your graph probably will not look exactly like **Figure 11-1.** Just try to generate a wave that has a constant amplitude.

24. Have your lab partner put the open string slightly out of tune again. Use only your ears to tune the open string. Adjust the pitch to match the fretted note. When you think you have tuned the string correctly, repeat step 21.

25. When you have finished, put away all of your materials, clean your work area, and wash your hands thoroughly.

▶ Analyzing Your Results

1. Calculate the elapsed times for sounds 1 and 2 in seconds by using the following equation. Record your answers in **Table 1.**

elapsed time = time of last peak − time of first peak

2. Calculate the elapsed time for sound 3 by using the following equation. Record your answer in **Table 2.**

elapsed time = time of last maximum − time of first maximum

3. Calculate the periods for sounds 1 and 2 in seconds by using the following equation. Record your answers in **Table 1.**

$$\textbf{period} = \frac{\textbf{elapsed time}}{\textbf{number of waves between first and last peaks}}$$

4. Calculate the beat period for sound 3 by using the following equation. Record your answer in **Table 2.**

$$\textbf{period} = \frac{\textbf{elapsed time}}{\textbf{number of cycles between first and last maximums}}$$

5. Calculate the frequency in Hertz for all three sounds by using the following equation. Record your answers for sounds 1 and 2 in **Table 1** and your answer for sound 3 in **Table 2.**

$$\textbf{frequency} = \frac{\textbf{1}}{\textbf{period}}$$

▶ Reaching Conclusions

6. Use your data for sounds 1 and 2 to describe the relationship between the pitch of a sound and the frequency of the wave that is generated.

7. Use the graph you printed or sketched for sound 3 to identify times when the two sound waves showed constructive interference. Why is the amplitude of the combined sound wave greater at these times than it is for each individual sound wave?

8. Use the same graph to identify times when the two sound waves showed destructive interference. Why is the amplitude of the combined sound wave smaller at these times than it is for each individual sound wave?

9. Which method of tuning an instrument worked better for you, using the microphone and graphing calculator or simply listening for differences in pitches? Explain why.

▶ Defending Your Conclusions

10. Musicians do not graph sound waves when they tune their instruments electronically. The electronic device they use tells them only whether the pitch is sharp (too high) or flat (too low). Does that make the results of your experiment invalid? Explain why or why not.

▶ Expanding Your Knowledge

1. Interview a local piano tuner, and ask the tuner to demonstrate the equipment used to tune pianos.

LAB

12 LIGHT

Choosing a Pair of Sunglasses

MATERIALS

- ✓ light sensor
- ✓ CBL
- ✓ TI graphing calculator and black link cable
- ✓ desk lamp
- ✓ piece of glass or aluminum foil
- ✓ ring stand
- ✓ clamp
- ✓ several different pairs of sunglasses, some having polarized lenses
- ✓ graph paper (optional)

Wear safety goggles and gloves while performing this experiment.

If you are working in the laboratory, wear a laboratory apron.

Avoid touching the light source during this experiment. A lit light bulb and its surroundings can become very hot.

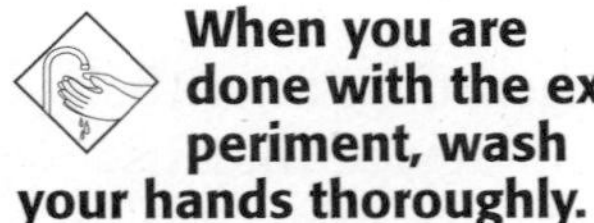

When you are done with the experiment, wash your hands thoroughly.

▶ Objectives

- ***Measure*** the intensity of light that is reflected from a smooth, flat surface.
- ***Compare*** several pairs of sunglasses for their ability to block reflected light while still allowing you to see well.
- ***Evaluate*** your data to determine which pair of sunglasses you would recommend that people wear.

▶ Identifying the Problem

Very intense reflected sunlight, or glare, can sometimes make it difficult for you to see. Wearing sunglasses is one way to protect your eyes from such blinding light. There are many different sunglasses to choose from, so how do you decide which pair to buy? Imagine that a consumer magazine has hired you to compare several different pairs of sunglasses for their ability to block reflected light. You must also determine how well you can see while wearing each pair.

▶ Finding Out More Information

If light is reflected from a rough surface, the light is reflected in many different directions. But light is reflected all in the same direction from a smooth surface, such as water, glass, or a flat piece of metal. The result is an unpleasant glare if the light is directed toward your eyes.

An effective pair of sunglasses is able to reduce the intensity of reflected light while still letting enough light pass through the lenses for you to see well. Sunglass manufacturers either add colors to the lenses, coat the lenses with reflective paints, or polarize the lenses (meaning that they make the lenses so that only light traveling in a certain direction can pass through).

▶ Coming Up with a Plan

The first thing you must do is simulate reflected sunlight. You can do this by setting up a desk lamp so that it reflects off glass or some other smooth surface. Then you can measure the intensity of the reflected light using a light sensor.

Next you will place the sunglasses between the reflected light and the sensor. This will allow you to measure how much of the reflected light is able to pass through the lenses of each pair of sunglasses.

You can determine whether the lenses are polarized by comparing how much light passes through them when they are positioned straight and then rotated. If the lenses aren't polarized, the intensity of the light passing through them will be about the same whether the lenses are rotated or not. If the lenses are polarized, the intensity of the light passing through them will be different when they are rotated.

▶ Performing the Experiment

Preparing for the experiment

1. Prepare a table in your lab report similar to the one shown below to record your data.

Sunglass Data

Light intensity reading without sunglasses	Brand/ style of sunglasses	Light intensity reading with sun-glasses horizontal	Light intensity reading with sun-glasses vertical	Maximum difference in light intensity readings
		USE YOUR DATASHEET		

Setting up the CBL System

2. Plug the light sensor into the Channel 1 input of the CBL. Connect the CBL to the graphing calculator by plugging the black link cable into the base of each unit.

3. Turn on both the CBL and the calculator. Press PRGM on the calculator, and select the PHYSCI program.

4. Go to the MAIN MENU, and select SET UP PROBES. Enter "1" as the number of probes. Select LIGHT from the SE-LECT PROBE menu. Enter "1" as the channel number.

5. From the MAIN MENU, select COLLECT DATA. Select MONITOR INPUT from the DATA COLLECTION menu. Light intensity readings will be displayed on the calculator.

Reflecting light

6. Set up the desk lamp, reflective surface, and light sensor as shown in **Figure 12-1. Do not place a pair of sunglasses in front of the light sensor yet.**

Figure 12-1

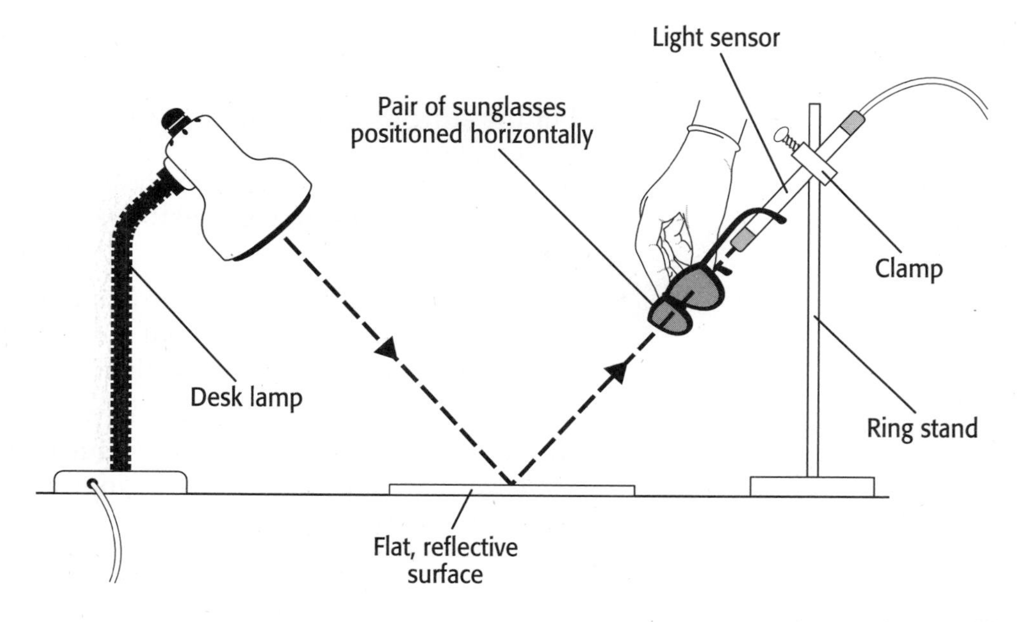

7. Adjust the lamp and light sensor so that the light sensor receives the maximum reflected light. Secure the light sensor with a clamp. You may also want to secure the desk lamp.

8. Without any sunglasses in front of the light sensor, record the intensity of the reflected light in your data table.

Testing the sunglasses

9. Hold a pair of sunglasses directly in front of the light sensor. Orient the glasses horizontally, as shown in **Figure 12-1** on the previous page. Record the intensity of the light in your data table.

10. Rotate the sunglasses 90 degrees, as shown in **Figure 12-2** below. Record the intensity of the light in your data table.

Figure 12-2

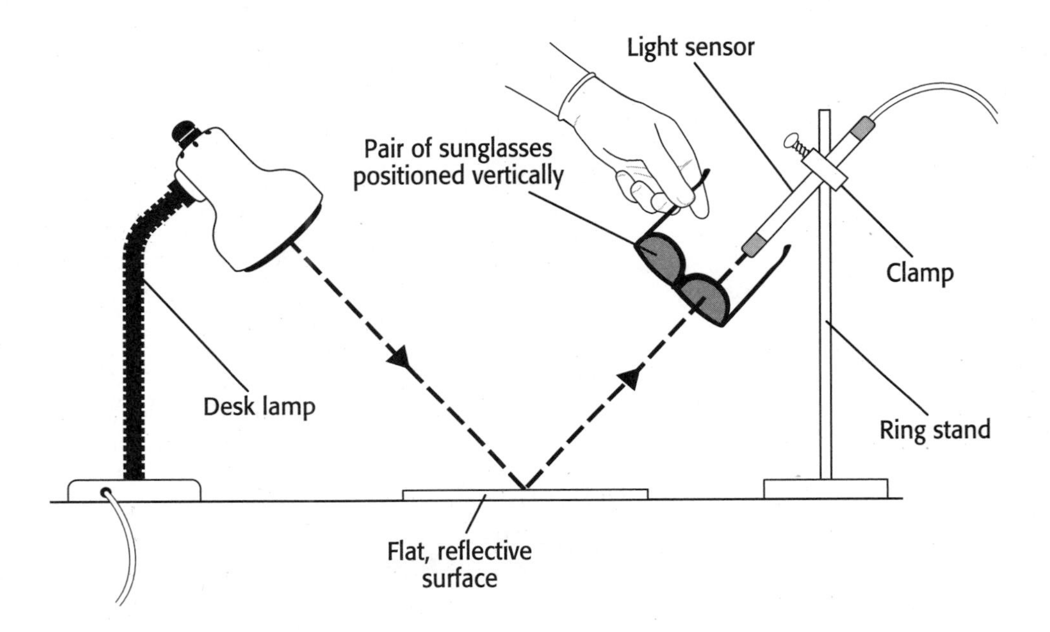

11. Note in your lab report what color the lenses are and any other important details about the sunglasses.

12. Put the sunglasses on, and look out a window at some reflected glare. (Do not look directly at the sun regardless of whether you are wearing sunglasses or not.) Then try reading some fine print in your textbook. Note in your lab report how well you can see.

13. Repeat steps 8–12 for each pair of sunglasses. Be sure to check the light intensity without sunglasses first each time to make sure the lighting conditions haven't changed.

14. When you have finished, press "+" on the calculator. Then turn off the light, put away all of your materials, and clean your work area. Be sure to wash your hands thoroughly before you leave.

▶ Analyzing Your Results

1. Calculate the maximum difference in light intensity for each pair of sunglasses by subtracting the smaller of the two measurements made with sunglasses from the measurement made without any sunglasses. Record your answers in your data table.

2. Plot your data on a bar graph like the one shown below. For each pair of sunglasses, plot the light intensity without any sunglasses (None), the light intensity with the sunglasses positioned horizontally (H), and the light intensity with the sunglasses positioned vertically (V).

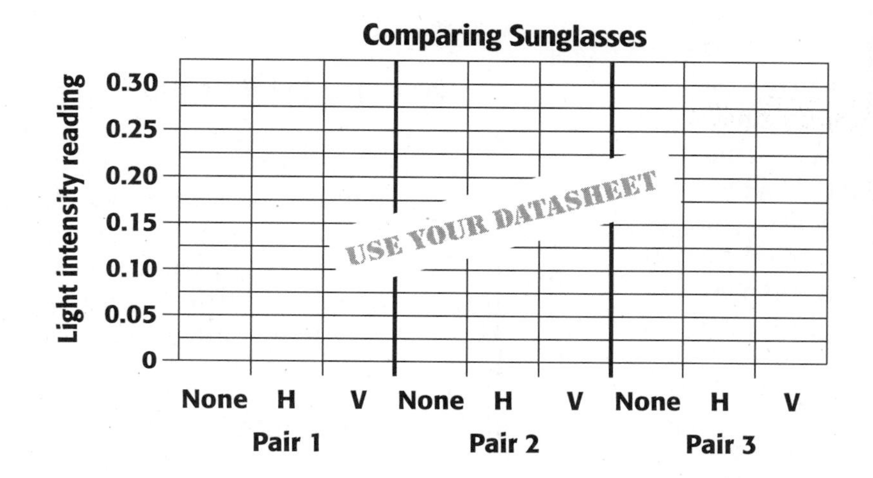

▶ Reaching Conclusions

3. Which pairs of sunglasses have polarized lenses? Could you read fine print very well while wearing these sunglasses? Explain why or why not.

4. Which sunglasses were able to reduce the reflected light intensity the most? Could you read fine print very well while wearing these sunglasses? Explain why or why not.

5. Of the sunglasses you tested, choose the pair that you think is the best overall. Explain your reasoning.

▶ Defending Your Conclusions

6. In this experiment, you tested only two of the many factors that make sunglasses effective. What are some other important factors in sunglass design?

▶ Expanding Your Knowledge

1. Perform a similar experiment to compare the intensity of light that is reflected from various flat surfaces. Try mirrors, water, bricks, light and dark paper, CD jewel cases, and anything else you have available.

Investigating How the Length of a Conductor Affects Resistance

MATERIALS

- ✓ voltage and current probes (and adapter cables)
- ✓ CBL
- ✓ TI graphing calculator and black link cable
- ✓ shallow plastic tray about 30 cm long
- ✓ 2% NaCl solution
- ✓ metric ruler
- ✓ dual channel amplifier
- ✓ 6 V battery
- ✓ 3 wires with alligator clips on each end
- ✓ 2 iron nails
- ✓ graph paper (optional)

Wear safety goggles, gloves, and a laboratory apron while performing this experiment.

Wires that are conducting electricity can become very hot. When collecting data, connect the battery only for short periods of time. Immediately disconnect the battery after you have collected data.

When you are done with the experiment, dispose of the waste as instructed by your teacher, and wash your hands thoroughly.

▸ Objectives

- ***Measure*** the current and voltage of a circuit several times, changing the length of one of the conductors in the circuit each time.
- ***Calculate*** the resistance of the circuit.
- ***Make a graph*** of your data to determine how increasing the length of one of the conductors in a circuit affects the resistance of the circuit.

▸ Identifying the Problem

For power companies that must transmit electrical energy over very long distances along power lines, knowing the resistance of the wires that transmit this energy is very important. If the resistance is known, one can calculate the power lost in transmission to each house—whether that house is one block or 80 km (50 mi) away from the power company. Imagine that you have been hired by the power company to determine how increasing the length of a conductor affects the resistance of the entire circuit.

▸ Finding Out More Information

For most conductors, the resistance of a circuit is related to voltage and current as described by Ohm's Law.

$$\textbf{current} = \frac{\textbf{voltage}}{\textbf{resistance}}$$

Voltage is measured in volts (V), resistance is measured in ohms (Ω), and current is measured in amperes (A). If the current and the voltage of a circuit are known, you can rearrange Ohm's Law to calculate the resistance, as shown below.

$$\textbf{resistance} = \frac{\textbf{voltage}}{\textbf{current}}$$

▸ Coming Up with a Plan

To examine how increasing the length of a conductor affects resistance, you have to be able to vary the conducting distance easily. What you can do is set up a circuit using a battery, three wires attached to alligator clips, and two iron nails sitting in a tray of salt water (NaCl solution). The salt water will complete the circuit by conducting electricity between the two nails.

To increase the conducting distance, you can move the nails farther apart. Each time you move the nails, you can measure the current and the voltage of the circuit. Then you can use these values to calculate the resistance of the circuit.

▶ Performing the Experiment

Preparing for the experiment

1. Prepare a table in your lab report similar to the one shown below to record your data.

Circuit Data

Distance between nails (cm)	Voltage (V)	Current (A)	Resistance (Ω)
2			
4			
8			
12			
16			
20			
24			
28			

USE YOUR DATASHEET

Setting up the CBL System

2. Plug the DIN 1 cable of the dual channel amplifier into the Channel 1 input of the CBL. Plug the DIN 2 cable into the Channel 2 input of the CBL. Plug the current probe into the PROBE 1 input of the dual channel amplifier. Plug the voltage probe into the PROBE 2 input of the dual channel amplifier.
3. Connect the CBL to the graphing calculator by plugging the black link cable into the base of each unit. Turn on both the CBL and the calculator. Press PRGM on the calculator, and select the PHYSCI program.
4. Go to the MAIN MENU, and select SET UP PROBES. Enter "2" as the number of probes. Select C-V CURRENT from the SELECT PROBE menu. Enter "1" as the channel number, and select USE STORED from the CALIBRATION menu.
5. With the current probe not connected to the circuit, press ENTER to zero the probe.
6. Select C-V VOLTAGE from the SELECT PROBE menu. Enter "2" as the channel number, and select USE STORED from the CALIBRATION menu.
7. Connect the two alligator clips from the voltage probe together. Press ENTER to zero the probe.
8. From the MAIN MENU, select COLLECT DATA. Select MONITOR INPUT from the DATA COLLECTION menu. Both current and voltage readings will be displayed on the calculator.

Assembling the circuit

9. Fill the shallow tray with a half-centimeter of 2% NaCl solution.

10. Use the figure below as a guide to assemble the circuit. All lines represent wires. Solid lines show the circuit you will be testing. Dotted lines show how current and voltage readings are transferred to the CBL and calculator.

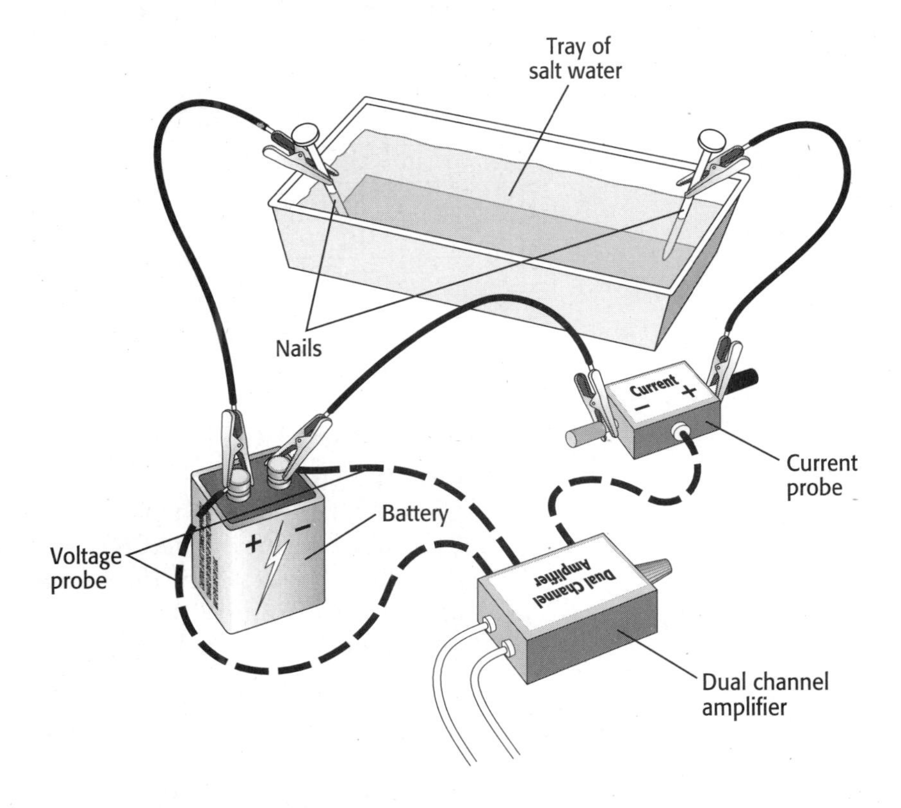

Measuring current and voltage

11. Use a metric ruler to place the two nails in the solution 2 cm apart. Record both the current and voltage in your data table.

12. Move the nails apart each distance listed in your data table. Record the current and voltage each time in the table.

13. When you have finished, press "+" on the calculator. Then put away all of your materials, clean your work area, and wash your hands thoroughly.

▶ Analyzing Your Results

1. Calculate the resistance of the circuit for each distance by using the following equation. Record your answers in your data table.

$$\textbf{resistance} = \frac{\textbf{voltage}}{\textbf{current}}$$

2. Plot your data on a graph like the one shown below. If you use your graphing calculator, be sure to include a sketch of the graph in your lab report.

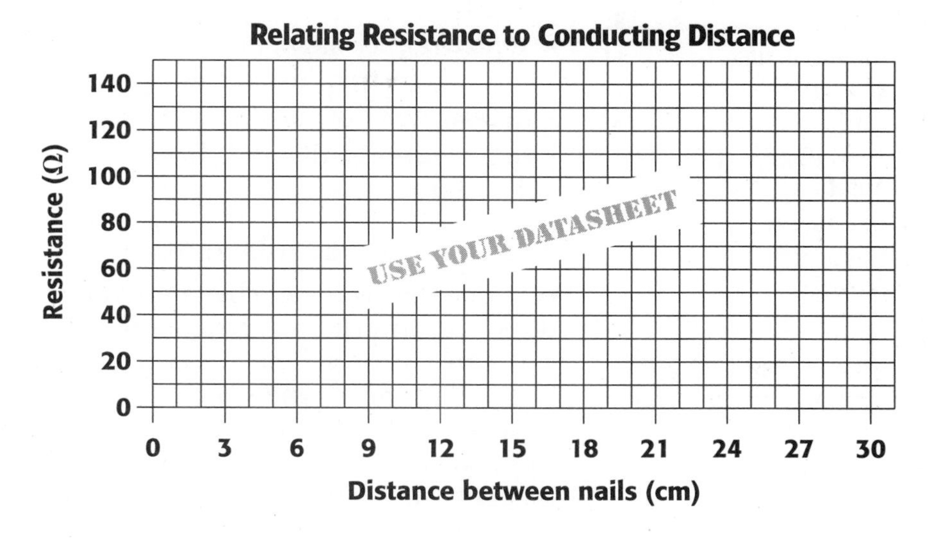

▶ Reaching Conclusions

3. What happens to the resistance of the circuit as the length of one of the conductors (the salt water) increases?
4. Both the salt water and the wires in the circuit are conductors. Which do you think conducts electricity more easily? Explain your answer.
5. Every conductor has some resistance. Compare the resistance of salt water with that of the wires in the circuit.
6. Suppose power is transmitted to your house from a transformer that is 45 m behind your house. The voltage supplied to your house is 110 V. If the power company guarantees a current of exactly 30 A at your breaker box (where the power enters your house), what is the maximum resistance per meter allowed in the power line to your house?

▶ Defending Your Conclusions

7. How would your results be different if you had varied the length of one of the wires in the circuit instead of the conducting distance of the salt water?

▶ Expanding Your Knowledge

1. Research the relative conductivity and cost of wires made from different metals. Summarize your findings in a poster that outlines the advantages and disadvantages of different types of wires.

LAB
14 MAGNETISM

Testing Magnets for an Electric Motor

MATERIALS

- ✓ magnetic field sensor (and adapter cable)
- ✓ CBL
- ✓ TI graphing calculator and black link cable
- ✓ bar magnet
- ✓ meterstick
- ✓ tape
- ✓ pencil
- ✓ graph paper (optional)

Wear safety goggles and gloves while performing this experiment.

If you are working in the laboratory, wear a laboratory apron.

When you are done with the experiment, wash your hands thoroughly.

▶ Objectives

- ***Measure*** the magnetic field strength of a bar magnet from several different distances.
- ***Investigate*** how distance affects magnetic field strength.
- ***Determine*** the distance from the magnet where the magnetic field is the strongest.

▶ Identifying the Problem

Imagine that you have been hired by a toy company to design a direct current (DC) electric motor for a small remote-controlled car. This motor will have a current-carrying coil that spins between two bar magnets. For the motor to work, the two magnets have to be close enough together to provide enough magnetic force to keep the coil spinning. To find out how far apart the magnets should be, you have decided to do an experiment to determine how magnetic force changes with distance.

▶ Finding Out More Information

A coil of wire with a current in it has a magnetic field. The coil is like a small magnet. When the coil is near a stronger magnet, the coil will move to align with the stronger magnet due to the magnetic force of the stronger magnet. In a motor, the current in the coil switches direction over and over again, so the coil keeps spinning as it tries to align with the magnetic field.

▶ Coming Up with a Plan

To determine how magnetic force is affected by distance, you can use a magnetic field sensor to measure the magnetic field strength of a magnet from several different distances. The SI unit for magnetic field strength is the gauss, or G.

You can either keep the magnet still and vary the placement of the magnetic field sensor, or you can keep the magnetic field sensor still and vary the placement of the magnet. If you vary the placement of the magnetic field sensor, Earth's magnetic field as well as the magnetic fields of any metal objects nearby may affect your measurements differently each time.

That's why it is better to keep the magnetic field sensor still and move only the magnet. Even if your measurements are affected by other magnetic fields, each of your measurements will be affected in the same way.

▶ Performing the Experiment

Preparing for the experiment

1. Prepare a table in your lab report similar to the one shown on the next page to record your data.

Magnetic Field Strength Data

Position of center of magnet (cm)	Magnetic field strength (G)	Position of center of magnet (cm)	Magnetic field strength (G)
0		15	
1		16	
2		17	
3		18	
4		19	
5		20	
6		21	
7		22	
8		23	
9		24	
10		25	
11		26	
12		27	
13		28	
14		29	

USE YOUR DATASHEET

Setting up your equipment

2. Tape the meterstick to the top of the table.
3. Tape the magnetic field sensor to the tabletop so that the white spot is facing up and is centered on the 14 cm mark. The sensor should be parallel to the tabletop and perpendicular to the meterstick, as shown below. To position the sensor that way, you may need to place a piece of cardboard underneath it. Make sure you do not move the sensor during the experiment.

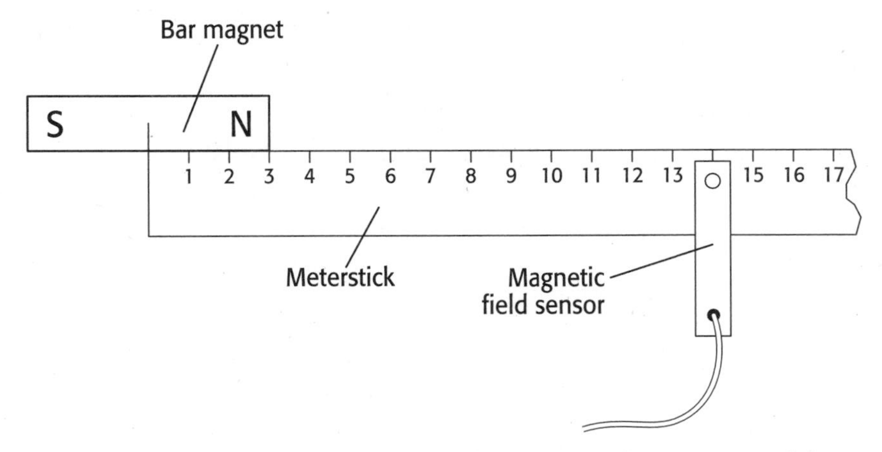

4. Use a pencil to make a small mark as close as possible to the center of the magnet.

Setting up the CBL System

5. Plug the magnetic field sensor into the Channel 1 input of the CBL. Connect the CBL to the graphing calculator by plugging the black link cable into the base of each unit. Set the switch on the sensor to "low."
6. Turn on both the CBL and the calculator. Press PRGM on the calculator, and select the PHYSCI program.
7. Go to the MAIN MENU, and select SET UP PROBES. Enter "1" as the number of probes. Select MAGNETIC FIELD from the SELECT PROBE menu, and enter "1" as the channel number.
8. Select USE STORED from the CALIBRATION menu. Select LOW (GAUSS) from the MG FIELD SETTING menu.
9. With the magnet very far away from the sensor, press ENTER to zero the sensor.
10. From the MAIN MENU, select COLLECT DATA. Select MONITOR INPUT from the DATA COLLECTION menu. Magnetic field readings will be displayed on the calculator.

Measuring magnetic field strength

11. Place the center of the bar magnet (where your pencil mark is) at the 0 cm mark on the meterstick. The N pole of the magnet should be facing toward the sensor, as shown on the previous page. Record the magnetic field strength in your data table.
12. Move the center of the magnet to the 1 cm mark. Record the magnetic field strength in your data table.
13. Move the center of the magnet to each centimeter mark listed in the table. Each time, record the magnetic field strength in your data table.
14. When you have finished, press "+" on the calculator, put away all of your materials, clean your work area, and wash your hands thoroughly.

▶ Analyzing Your Results

1. Plot your data on a graph like the one shown on the next page. If you use your graphing calculator, be sure to include a sketch of the graph in your lab report.
2. Describe the relationship shown by the graph. What happens as the N pole of the magnet gets closer to the magnetic field sensor? What happens as the S pole of the magnet gets farther away from the magnetic field sensor?
3. Where is the center of the magnet when the magnetic field strength equals zero? Explain why.

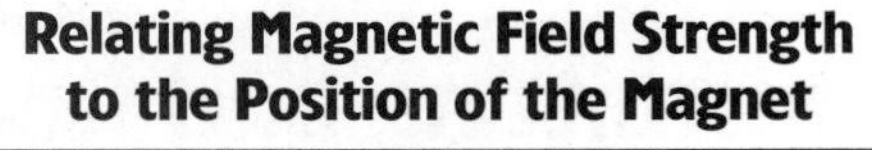

Relating Magnetic Field Strength
to the Position of the Magnet

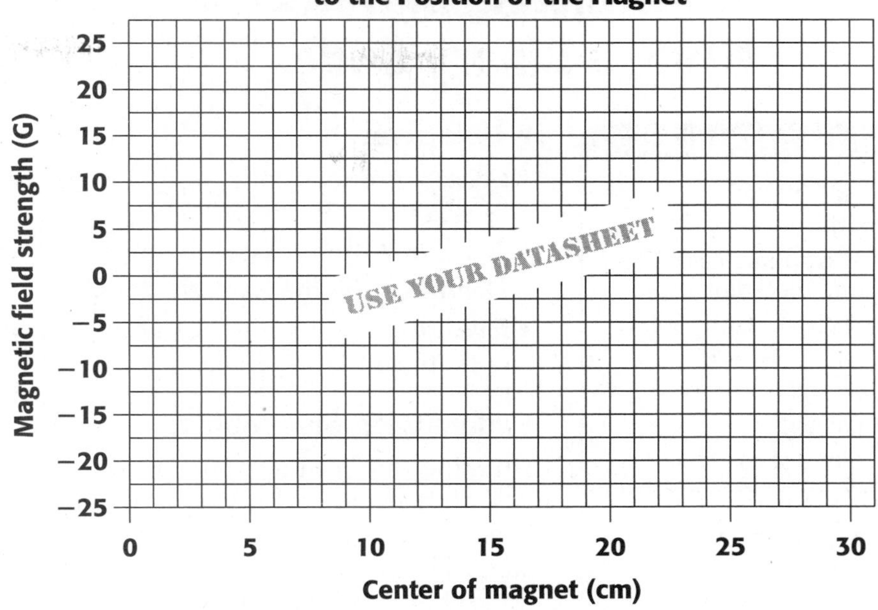

4. How would your graph look different if in step 11 you had placed the bar magnet so that its S pole was facing toward the magnetic field sensor?

▶ Reaching Conclusions

5. Where is the center of the magnet when the magnetic field is the strongest? How far away is the end of the magnet from the sensor?

6. If you designed the motor for a toy car using two magnets similar to those used in this experiment, how much space would you leave between them to generate the strongest possible magnetic field? (**Hint:** You want each magnet to have the strongest possible magnetic field.)

▶ Defending Your Conclusions

7. The electric motor you are designing will have two magnets. Explain why you were able to determine how far apart the two magnets should be placed by measuring the magnetic field strength of only one magnet.

▶ Expanding Your Knowledge

1. Use the magnetic field sensor to investigate the properties of an electromagnet. Try changing the wire size, the number of turns, and the core of the electromagnet to find out how each of these factors affects the electromagnet's field strength.

LAB 15 COMMUNICATION TECHNOLOGY

Transmitting and Receiving a Message Using a Binary Code

MATERIALS

- ✓ light sensor
- ✓ CBL
- ✓ TI graphing calculator and black link cable
- ✓ TI-Graph Link cable
- ✓ computer with a printer
- ✓ flashlight
- ✓ piece of cardboard

Wear safety goggles while performing this experiment.

If you are working in the laboratory, wear gloves and a laboratory apron.

When you are done with the experiment, wash your hands thoroughly.

▶ Objectives

- ***Create*** a binary code, and use the code to write a short message.
- ***Transmit*** the message using a flashlight.
- ***Decipher*** the message that is sent.
- ***Evaluate*** the advantages and disadvantages of using a binary code to send a message.

▶ Identifying the Problem

Imagine that an international team of scientists has developed a research station with two goals in mind. The first goal is to monitor space and search for any signals that may contain a message. The second goal is to develop a simple but clear binary code that can be used to send messages to any intelligent lifeforms that might exist in space. Imagine that your team has been asked to help develop this binary code. Your team must then test the code you developed by determining how easy it is to transmit and receive a message using the code.

▶ Finding Out More Information

A binary code is digital because it uses only two values to represent all of the possible symbols that may be transmitted. For example, sending a signal with a flashlight is a form of binary code. The two possible values are "on" (light is on) and "off" (light is off). Computers also work by using a binary code in which "on" = 1 and "off" = 0.

This code can be translated into a digital signal. The graph below shows that a digital signal is not smooth and continuous like most of the graphs you have encountered so far. That's because when the value changes, it changes abruptly, not gradually. For example, a flashlight is either on or off, never in between. The idea is similar for the binary code of a computer.

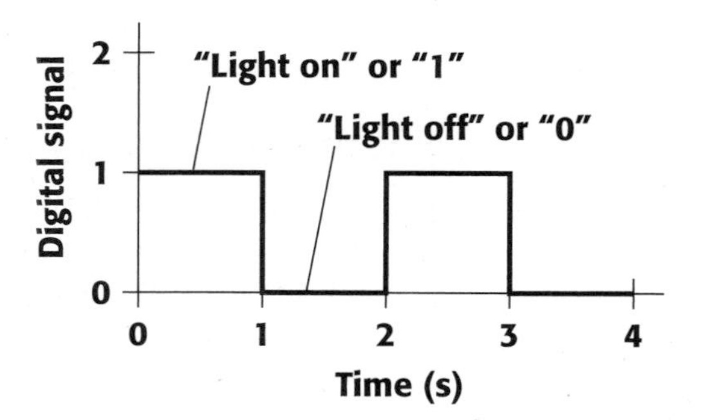

▶ Coming Up with a Plan

A binary code uses only two values that are often referred to as bits—the numbers 1 and 0. You will have to use combinations of these bits to represent different symbols. The table below shows that if you increase the number of bits used to represent each symbol, you have more possibilities for your message. For example, if you represent each symbol with 5 bits, your message can have all 26 letters of the alphabet.

The Effect of the Number of Bits on Your Binary Code

Number of bits for each symbol	Total number of symbols possible	Binary code examples
1	2	0, 1
2	4	00, 01, 10, 11
3	8	000, 001, 010, 011, 100, 101, 110, 111
4	16	0000, 0001, . . . , 1110, 1111
5	32	00000, 00001, . . . , 11110, 11111

Instead of turning the flashlight on and off to send your message, you can leave the switch turned on and turn the light "off" by covering it with a piece of cardboard. The CBL System will let you receive the message quickly, store it, and then convert it to a graph for easier interpretation.

▶ Performing the Experiment

Preparing for the experiment

1. Determine the number of bits you will need for each symbol to represent all of the symbols in your message. Make a table similar to the one shown below, and complete it by assigning each symbol a binary code. You can either make a code similar to this one or you can make a different code.

Your Group's Binary Code

Symbol	Binary code
a	00000
b	00001
etc.	etc.

2. One member of your group should decide on a short, one-word message to send. This person should write the message on a piece of paper according to the binary code your group created. The rest of the group should continue on to step 3.

Setting up the CBL System

3. Plug the light sensor into the Channel 1 input of the CBL. Connect the CBL to the graphing calculator by plugging the black link cable into the base of each unit.
4. Turn on both the CBL and the calculator. Press PRGM on the calculator, and select the PHYSCI program.
5. Go to the MAIN MENU, and select SET UP PROBES. Enter "1" as the number of probes. Select LIGHT from the SELECT PROBE menu. Enter "1" as the channel number.
6. From the MAIN MENU, select COLLECT DATA. Select TIME GRAPH from the DATA COLLECTION menu. Enter "0.05" as the time between samples in seconds. Enter "99" as the number of samples. You will have 5 s to transmit the bits that represent a symbol.
7. Select USE TIME SETUP. The calculator screen should say READY EQUIPMENT. You'll have to transmit the message one letter or number at a time because the memory of the graphing calculator is limited to a 5 s interval.

Transmitting the message

8. Position the flashlight and light sensor on a table so that the flashlight points directly toward the light sensor. The flashlight should be about 0.5 m (1.6 ft) from the sensor.
9. When you are ready to transmit the first symbol of the message, turn the flashlight on. Practice moving the piece of cardboard back and forth quickly to cover and uncover the light. You might find it helpful to say what you are signaling as you practice (for example, "on, on, off, on, off"). Each value (on or off) should be uniform in length and should last about 1 s.
10. After you have practiced, cover the light with the piece of cardboard. Press ENTER on the calculator, and immediately send the appropriate signal for the first symbol of your message.
11. When the CBL is done collecting data, press ENTER. A graph will appear on the screen. Either make a sketch of the graph in your lab report or disconnect the calculator from the CBL and connect it to a computer using the TI-Graph Link cable. Use the computer to save and print the graph.
12. Reconnect the calculator to the CBL. Press ENTER, and select YES when asked if you want to repeat.
13. Repeat steps 10–12 for each symbol of your message.
14. When you have finished sending the message, put away all of your materials, and clean your work area. Give the graphs and the piece of paper with the message written in binary code to your lab partners so that they can decipher the message. Be sure to wash your hands before you leave.

▶ Analyzing Your Results

1. Interpret the graphs by writing the binary code for each symbol using 1's and 0's. Check with your lab partner to find out if your interpretation was right. Correct any errors you may have made.
2. What message did your lab partner send?

▶ Reaching Conclusions

3. Compare the binary code you created with any other codes you have used. What are the advantages of using a binary code to send a message? What are the disadvantages?
4. How could you change your code to include capital letters and punctuation marks?
5. The calculator can be programmed with a minimum sampling time of 0.000 164 s. How many symbols could you send in 1 minute using this sampling time? (**Hint:** First determine how many bits of information could be sent in 1 minute, and then divide this number by the number of bits you used to represent each symbol.)
6. How could an analog signal be sent using a light source? How would the graph of an analog signal differ from the ones you generated in this experiment?

▶ Defending Your Conclusions

7. Intelligent life-forms in space would most likely not speak or understand the language your message was written in. Describe a message you could send that might be universally understood.

▶ Expanding Your Knowledge

1. Research the ASCII (American Standard Code of Information Interchange) code used in computers. Describe the code itself and how it is used in computers today.
2. There are several scientific research projects seeking intelligent life-forms in the universe with the use of radio telescopes. Create a poster that describes the goals, methods, and research activities of one of these projects. Two possible projects include the SETI (Search for Extraterrestrial Intelligence) and SERENDIP (Search for Extraterrestrial Radio Emissions from Nearby Developed Intelligent Populations).

LAB

16 THE UNIVERSE

Determining the Speed of an Orbiting Moon

MATERIALS

- ✓ force sensor (and adapter cable)
- ✓ CBL
- ✓ TI graphing calculator and black link cable
- ✓ 200 – 300 g mass
- ✓ stopwatch
- ✓ 25 lb fishing line that is 1.0 m long
- ✓ swivel snap
- ✓ small plastic tube
- ✓ meterstick
- ✓ marker
- ✓ graph paper (optional)
- ✓ calculator (optional)

Wear safety goggles and gloves while performing this experiment.

If you are working in the laboratory, wear a laboratory apron.

This experiment involves whirling a small mass around in circles over your head. Make sure that you have enough room to collect data without hitting anyone or anything.

When you are done with the experiment, wash your hands thoroughly.

▶ Objectives

- ***Make several models*** of a moon orbiting a planet, each time modeling a planet of a different mass.
- ***Determine*** the distance each "moon" travels in 20 revolutions.
- ***Calculate*** the speed of each "moon," and relate it to the mass of the planet being modeled.

▶ Identifying the Problem

A media production company is producing an educational film that will show an imaginary solar system. In the opening animation, there will be a slow-motion view of a solar system with planets orbiting the sun and moons orbiting the planets.

Imagine that the producer has asked you to help develop the animation. He wants to be sure that the moons orbit their planets at the proper relative speed. You have been asked to develop a model that will help predict the speeds at which the moons should orbit the planets.

▶ Finding Out More Information

When a moon orbits a planet, its speed is determined by the gravitational force between the moon and the planet. This force is affected by the mass of the planet as well as the distance between the moon and the planet. The greater the mass of the planet, the greater the force between the moon and planet. The greater the distance between the moon and the planet, the smaller the force between them.

▶ Coming Up with a Plan

To make your models, you can attach a small mass to one end of a fishing line. Then you can swing the fishing line with the attached mass in circles at a constant speed, as shown on the next page. When you do this, the force exerted on the fishing line will represent the force of gravity between the "moon" and "planet." You can measure this force using a force sensor. You need to determine how the speed changes as you change the force of gravity.

▶ Performing the Experiment

Preparing for the experiment

1. Prepare a table in your lab report similar to the one shown on the next page to record your data.
2. Securely attach the swivel snap to one end of the fishing line by tying a knot. Feed the line through the plastic tube, and securely connect the swivel snap to the force sensor. Securely connect the other end of the line to the mass by tying another knot.

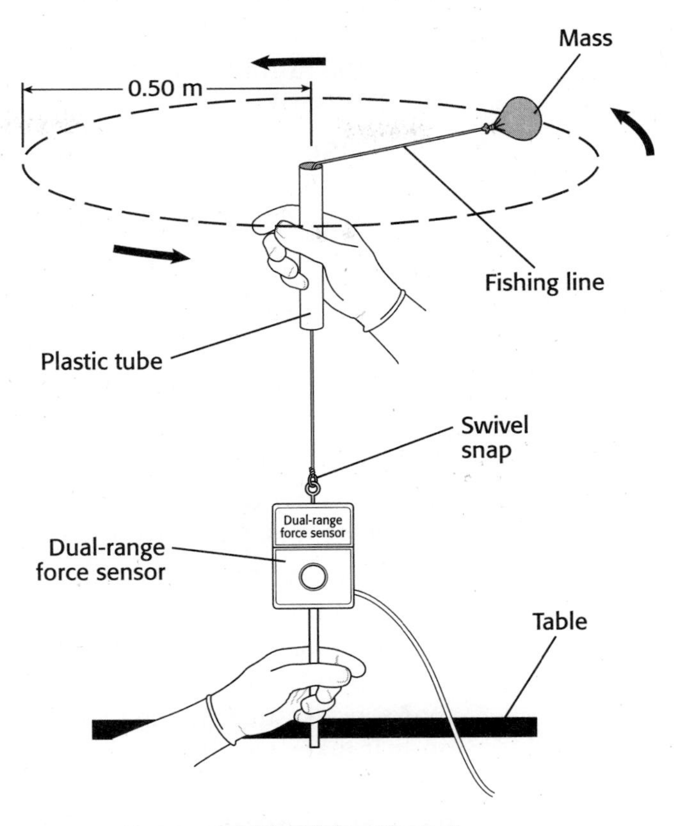

Forces and Orbiting Speeds

Force (N)	Time to complete 20 revolutions (s)	Distance traveled in 20 revolutions (m)	Orbiting speed (m/s)
5.0			
7.5			
10.0			
12.5			
15.0			
17.5			
20.0			

USE YOUR DATASHEET

3. Use a marker to mark the fishing line 0.50 m away from the mass. When you are swinging the mass, keep this mark even with the top of the plastic tube.

Setting up the CBL System

4. Plug the force sensor into the Channel 1 input of the CBL. Connect the CBL to the graphing calculator by plugging the black link cable into the base of each unit.

5. Turn on both the CBL and the calculator. Press PRGM on the calculator, and select the PHYSCI program.

6. Go to the MAIN MENU, and select SET UP PROBES. Enter "1" as the number of probes. Select FORCE from the SELECT PROBE menu. Enter "1" as the channel number.
7. Select USE STORED from the CALIBRATION menu. Select DUAL-RANGE from the FORCE PROBE menu, then select DUAL-RANGE 50 N. Set the switch on the sensor to 50 N.
8. From the MAIN MENU, select COLLECT DATA. Select MONITOR INPUT from the DATA COLLECTION menu. Force readings will be displayed on the calculator.

Varying force by changing the orbiting speed of the mass

9. Secure the force sensor to a tabletop by holding it firmly with one hand. Have someone else in your group grip the plastic tube. Adjust the position of the plastic tube so that the top of it is even with the mark on the fishing line.
10. The person holding the plastic tube should then start the mass swinging. Another person should monitor the force reading. Speed up or slow down the speed of the orbiting mass to maintain a force of 5.0 N. When the reading remains constant, use the stopwatch to time how many seconds it takes for the mass to complete 20 revolutions. Record the time in your data table.
11. Repeat step 10, this time maintaining a force of 7.5 N. Record the time it takes for the mass to complete 20 revolutions in your data table.
12. Repeat step 10 five more times, maintaining the following forces: 10.0 N, 12.5 N, 15.0 N, 17.5 N, and 20.0 N. Record the time it takes for the mass to complete 20 revolutions each time in your data table.

▶ Analyzing Your Results

1. Calculate the distance the mass travels in meters in one revolution by using the following equation. The radius is equal to 0.50 m.

$$\textbf{distance} = 2\pi \times \textbf{radius}$$

   Multiply this distance by 20 to determine the total distance the mass traveled for each trial. Record your answer in your data table.

2. For each force, calculate the speed of the mass by using the following equation. Record your answers in your data table.

$$\textbf{speed} = \frac{\textbf{distance traveled in 20 revolutions}}{\textbf{time to complete 20 revolutions}}$$

3. Plot your data on a graph like the one shown below. If you use your graphing calculator, be sure to include a sketch of the graph in your lab report.

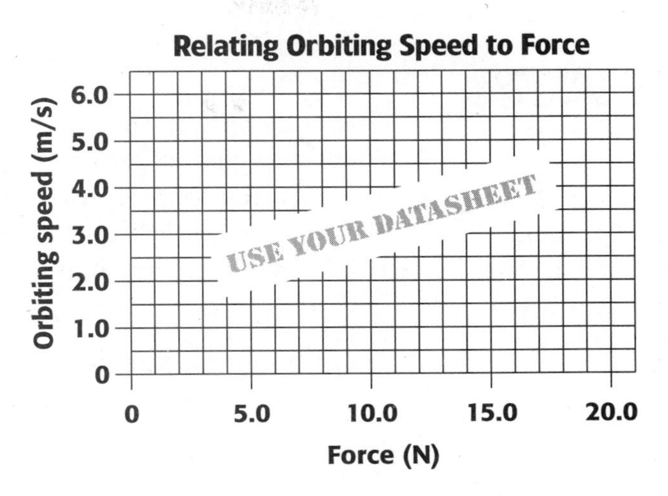

▶ Reaching Conclusions

4. Use your graph to describe the relationship that exists between the orbiting speed of an object and the force acting on it.

5. Suppose that the opening animation of the film will show two planets. One of the planets will be twice as massive as the other planet. Compare the speeds of the moons orbiting these planets.

6. You can determine the mass of a planet if you know the distance between the planet and one of its moons and the orbiting speed of the moon, as shown by the following equation.

$$\text{mass of planet} = (\text{speed})^2 \times \text{radius} \times \left(1.499 \times 10^{10}\ \frac{\text{kg} \cdot \text{s}^2}{\text{m}^3}\right)$$

Suppose a planet and its moon are 250 000 000 m apart, and the speed of the orbiting moon is 320 m/s. Use the equation above to determine the mass of the planet in kilograms.

▶ Defending Your Conclusions

7. Suppose someone tells you that your results are not valid because the mass and forces you measured in this experiment were too small to be comparable to a planet and its moon. How could you show that your results are valid?

▶ Expanding Your Knowledge

1. Design an experiment similar to the one you just did to determine how changing the distance (the radius) would affect the orbiting speed of a moon. Perform the experiment, and summarize your findings in a report.

LAB 17 PLANET EARTH

Relating Convection to the Movement of Tectonic Plates

MATERIALS

- ✓ temperature probe
- ✓ CBL
- ✓ TI graphing calculator and black link cable
- ✓ aquarium or large tank
- ✓ small jar with a hole in the lid
- ✓ food coloring
- ✓ cold tap water
- ✓ hot tap water
- ✓ hot pad
- ✓ tongs
- ✓ stirring rod
- ✓ Teflon® tape
- ✓ stopwatch
- ✓ graph paper (optional)

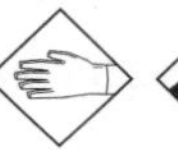

Wear safety goggles, gloves, and a laboratory apron while performing this experiment.

Use caution when handling hot water. Use tongs to transfer the jar filled with hot water into the aquarium.

When you are done with the experiment, dispose of the waste as instructed by your teacher, and wash your hands thoroughly.

▶ Objectives

- ***Observe*** convection currents using an aquarium filled with cold water and a small jar filled with hot water and food coloring.
- ***Measure*** the temperature changes of four different areas in the aquarium over time.
- ***Evaluate*** the theory that convection currents cause tectonic plates to move.

▶ Identifying the Problem

Large tectonic plates, which are pieces of Earth's lithosphere, move from 1–16 cm (0.4–6 in.) each year. This may not seem like much, but over millions of years this movement results in a dramatic change of Earth's surface.

Nobody knows exactly why tectonic plates move. One theory claims that leftover heat from the formation of Earth causes convection currents in the asthenosphere. Scientists who favor this theory believe that these currents of rock might push the plates, causing them to move. Imagine that you have been asked to perform an experiment to simulate these conditions to determine if this theory is valid.

▶ Finding Out More Information

Convection is one of the ways energy is transferred as heat. When a substance is heated, it often becomes less dense. As a result, the hot substance is more buoyant than its surroundings, causing it to rise. After rising, the substance cools and returns to its original density. This causes the substance to sink back down. If the substance gets hot again, the cycle repeats.

Over time, a pattern of moving material—a convection current—is established. The pattern is usually one in which hot material rises in the center, spreads out, and then cools. When the material cools, the part that spreads the farthest sinks back down.

▶ Coming Up with a Plan

To model Earth, you can use a large aquarium filled with cold water. You can use a small jar filled with hot water to act as a hot spot in the asthenosphere. The jar needs to have a hole in the lid so that the hot water inside can come out. You will have to add food coloring to the hot water so that you can see how the "lava" flows.

You will also need to determine if the water in the aquarium is moving. If it is moving, you can determine which direction it is moving by comparing the temperatures of four different stations in the aquarium over a 10-minute period.

▸ Performing the Experiment

Preparing for the experiment

1. Prepare a table in your lab report similar to the one shown below to record your data.

Temperature Readings and Observations

Time (min)	Temperature (°C)				Observations
	Station 1	Station 2	Station 3	Station 4	
0					
0.5					
1					
3					
5					
10					

USE YOUR DATASHEET

2. Divide your group into two teams. While one team is preparing the aquarium, the other team should move on to step 5.
3. Fill the aquarium with cold tap water.
4. Place five drops of food coloring in a small jar. Set the jar aside until the other team has finished setting up the CBL System.

Setting up the CBL System

5. Plug the temperature probe into the Channel 1 input of the CBL. Connect the CBL to the graphing calculator by plugging the black link cable into the base of each unit. Use Teflon tape to attach the temperature probe to one end of the stirring rod, as shown on the next page.
6. Turn on both the CBL and the calculator. Press PRGM on the calculator, and select the PHYSCI program.
7. Go to the MAIN MENU, and select SET UP PROBES. Enter "1" as the number of probes. Select TEMPERATURE from the SELECT PROBE menu, and enter "1" as the channel number.
8. From the MAIN MENU, select COLLECT DATA. Select MONITOR INPUT from the DATA COLLECTION menu. Temperature readings will be displayed on the calculator.

Monitoring convection currents

9. Holding the stirring rod, gently lower the temperature probe to station 1 of the aquarium, as shown on the next page. Wait for the reading to stabilize, and then record the temperature in your data table.

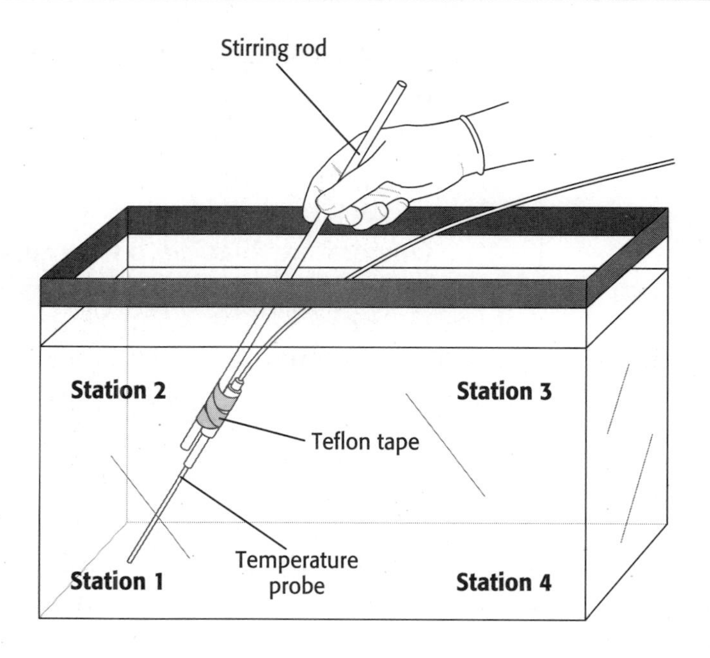

10. Without removing the temperature probe from the water, repeat step 9 for stations 2, 3, and 4. Then move the temperature probe back to station 1.
11. Fill the jar containing food coloring with hot tap water. Use tongs or a hot pad when handling this jar. Put the lid on the jar securely. Use tongs to gently lower the jar into the aquarium. Position the jar at station 4, and start the stopwatch.
12. After 30 s (0.5 minutes) have passed, record the temperature at station 1 in your data table.
13. Without removing the temperature probe from the water and being careful not to disturb the water too much, slowly move the probe to station 2. Once the reading has stabilized, record the temperature in your data table.
14. Repeat step 13 for stations 3 and 4. Record the temperature each time in your data table.
15. Without removing the temperature probe from the water, slowly move the probe back to station 1. Record any observations you make about the movement of the food coloring in your lab report.
16. After 1 minute has passed, record the temperature at station 1 in your data table. Then repeat steps 13–15.
17. Repeat step 16 after 3 minutes have passed, then after 5 minutes, and then after 10 minutes. Record the temperature of each station each time in your data table.
18. When you have finished, press "+" on the calculator. Then dispose of the waste as instructed by your teacher, put away all of your materials, and clean your work area. Be sure to wash your hands thoroughly before you leave.

▶ Analyzing Your Results

1. Plot your data on a graph like the one shown below. Plot the data for each station on the same set of axes, using a different color for each set of data. Be sure to label your graph clearly.

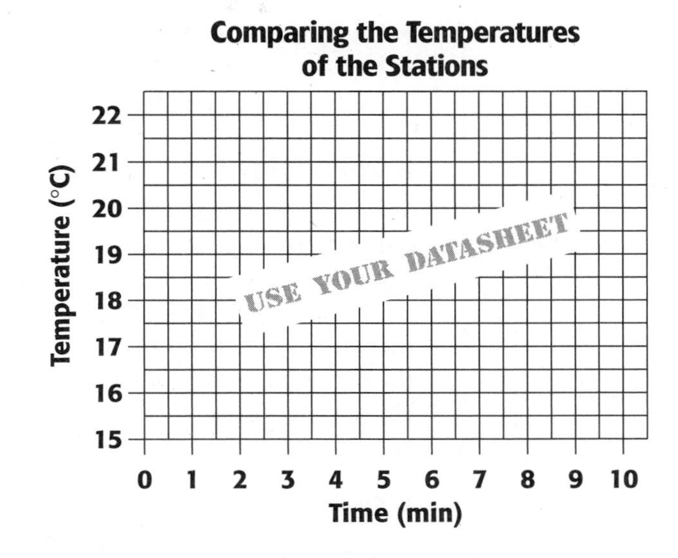

2. Which station showed the greatest temperature change over the 10-minute period? Which station's temperature changed the least over the 10-minute period?

3. Use your data and your graph to describe how energy was transferred as heat between stations in the aquarium.

▶ Reaching Conclusions

4. Describe the rate at which food coloring left the jar when temperature differences were large. What happened to the rate as the temperature differences were smaller?

5. Do your results support the theory that convection currents may be causing tectonic plates to move? Explain why or why not.

6. If no new heat is being generated inside Earth, what does this experiment predict will eventually happen?

▶ Defending Your Conclusions

7. To model convection currents in Earth's asthenosphere, you used an aquarium filled with cold water and a small jar filled with hot water. In what ways does Earth differ from your model? How could you change your model to make it more realistic?

▶ Expanding Your Knowledge

1. Research how the solar system was formed. Include the sources of heat, the rates of cooling, and the substances that were present. Create a poster that outlines your findings.

LAB

18 THE ATMOSPHERE

Predicting Coastal Winds

MATERIALS

- ✓ 2 temperature probes
- ✓ CBL
- ✓ TI graphing calculator and black link cable
- ✓ 150 mL beakers (2)
- ✓ tap water at room temperature
- ✓ sand
- ✓ reflector lamp with a clamp and a 300 W light bulb
- ✓ 2 ring stands
- ✓ 2 small clamps
- ✓ large rod

Wear safety goggles and gloves while performing this experiment.

If you are working in the laboratory, wear a laboratory apron.

Avoid touching the light source during this experiment. A lit light bulb and its surroundings can become very hot.

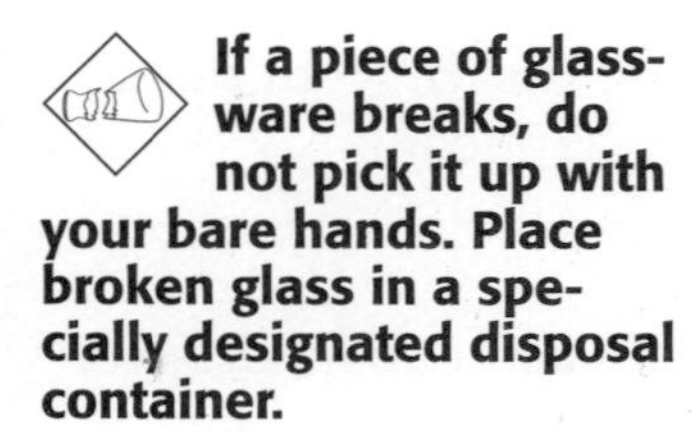

If a piece of glassware breaks, do not pick it up with your bare hands. Place broken glass in a specially designated disposal container.

When you are done with the experiment, wash your hands thoroughly.

▶ Objectives

- ***Make a model*** of a coast on a sunny day using a lamp, sand, and water.
- ***Compare*** the temperature of the sand with that of the water.
- ***Predict*** which way the wind is likely to be blowing near the coast on a typical sunny day.

▶ Identifying the Problem

When sailing a sailboat, you must be able to predict and take full advantage of the wind. A potentially dangerous situation may result if you lose control of the sailboat close to the shore when the wind is blowing toward the shore. In many coastal areas, the wind blows toward and away from the shore in predictable ways throughout the day. In this experiment, you will predict which way the wind is likely to be blowing on a typical sunny day.

▶ Finding Out More Information

The direction in which the local winds blow is influenced by Earth's rotation as well as the local terrain. Some of Earth's surfaces get hotter than other surfaces do as they absorb energy from the sun. Air above a hotter surface expands and rises. Air above a cooler surface nearby then flows in to replace the heated air, resulting in wind.

▶ Coming Up with a Plan

To find out which way the wind is likely to be blowing, you need to make a model of a coastal area on a sunny day. You can use a lamp to model the sun. To model the shore and ocean, you can use a beaker filled with sand and a beaker filled with water.

If you shine the lamp on both the sand and the water, you can measure how the temperature of each substance changes with time. You can then use this information to predict which way the wind is likely to be blowing near the shore.

▶ Performing the Experiment

Preparing for the experiment

1. Prepare a table in your lab report similar to the one shown on the next page to record your data.

Setting up the CBL System

2. Plug one of the temperature probes into the Channel 1 input of the CBL. Plug the other temperature probe into the Channel 2 input of the CBL. Connect the CBL to the graphing calculator by plugging the black link cable into the base of each unit.

Temperature Changes of Water and Sand

Time (min)	Temperature of water (°C)	Temperature of sand (°C)
0		
1		
2		
3		
4		
5		
6		
7		
8		
9		
10		
Total temperature change		

3. Turn on both the CBL and the calculator. Press PRGM on the calculator, and select the PHYSCI program.
4. Go to the MAIN MENU, and select SET UP PROBES. Enter "2" as the number of probes. Select TEMPERATURE from the SELECT PROBE menu. Enter "1" as the channel number. Select TEMPERATURE again from the SELECT PROBE menu, and enter "2" as the channel number.
5. From the MAIN MENU, select COLLECT DATA. Select TIME GRAPH from the DATA COLLECTION menu. Enter "60" as the time between samples in seconds. Enter "10" as the number of samples.

Setting up your model

6. Carefully fill one of the beakers with 100 mL of room-temperature tap water. Fill the other beaker with 100 mL of room-temperature sand.
7. Assemble the model as shown on the next page. **Do not turn the lamp on yet.** The lamp should be about 30.5 cm (12 in.) above the surface of the sand and water and should shine on both substances with equal intensity.
8. Position the Channel 1 temperature probe below the surface of the water so that it is in the center of the beaker, and clamp it into place. Make sure the probe is not touching the sides or bottom of the beaker.
9. Position the Channel 2 temperature probe below the surface of the sand in the same way, and clamp it into place. Again make sure the probe is not touching the beaker.

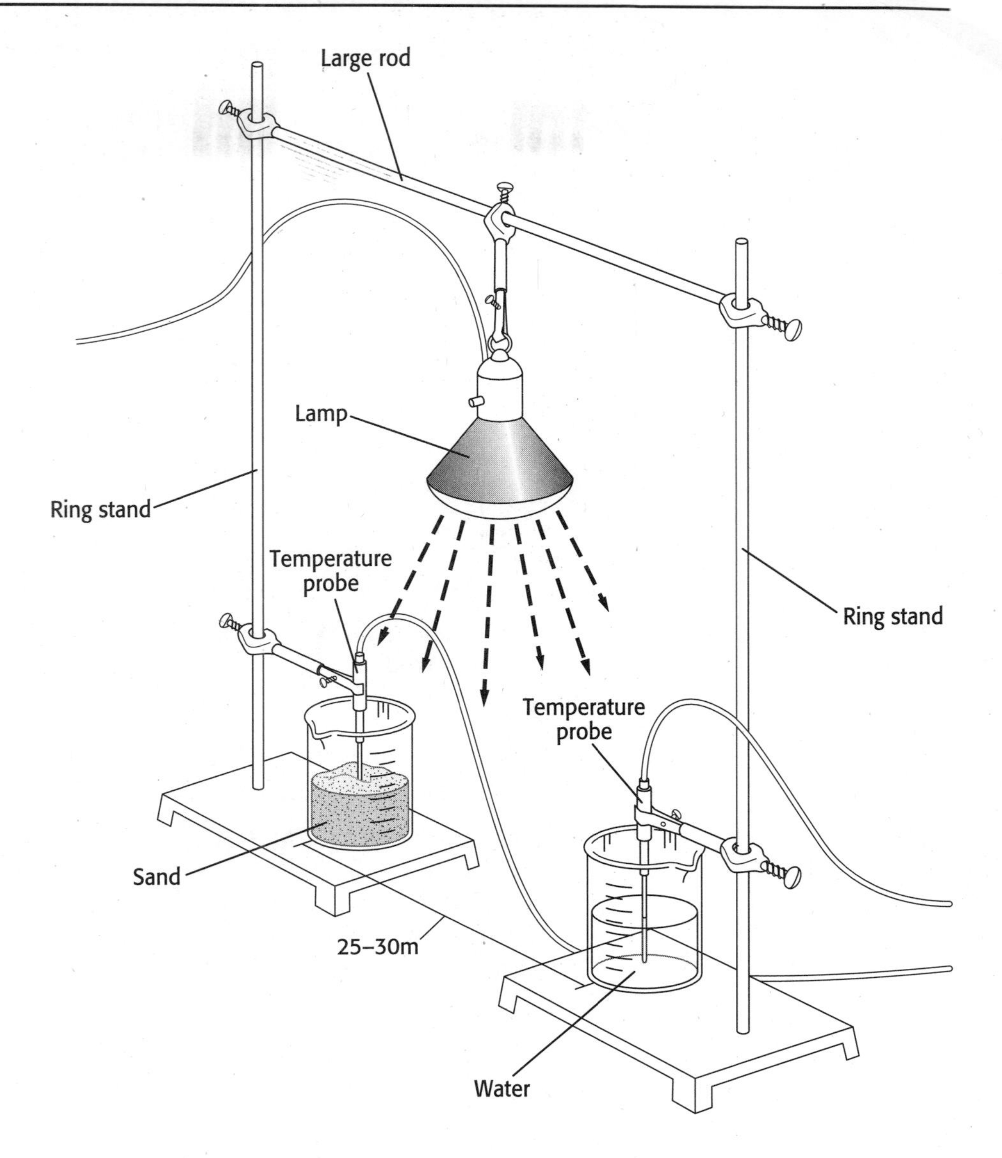

Measuring temperature changes

10. Select USE TIME SETUP. Enter "0" for the y-axis minimum. Enter "40" for the y-axis maximum. Enter "5" for the y-axis scale. Wait about 3 minutes for the probes to acclimate to the temperatures of the water and the sand. Then turn the lamp on, and press ENTER on the calculator.

11. When the CBL is done collecting data, press ENTER. A graph of how the temperature of the water changed with time will appear. Use the arrow keys to trace the points on the graph. Record the temperature of the water each minute in your data table.

12. Press ENTER again to see a graph of how the temperature of the sand changed with time. Again use the arrow keys to trace the points on the graph. Record the temperature of the sand each minute in your data table.

13. When you have finished, turn off the light. Then put away all of your materials, clean your work area, and wash your hands thoroughly.

▶ Analyzing Your Results

1. Calculate the total temperature change for both the water and the sand by subtracting the initial temperature of each substance (temperature at 0 minutes) from the final temperature of each (temperature at 10 minutes). Record your answers in your data table.

▶ Reaching Conclusions

2. Did the sand or the water have the greater temperature change during the 10-minute period?
3. Why was the temperature difference so large for this substance compared with the other substance?
4. Would you expect the winds to be blowing toward the shore or away from the shore in the middle of the day? Explain your reasoning.

▶ Defending Your Conclusions

5. Do you think your model of a coast is realistic? How does a real coastline differ from your model?

▶ Expanding Your Knowledge

1. Build and calibrate an instrument to measure the direction and speed of the wind at your school or at your home. Monitor the wind direction for several weeks, and try to develop a system for predicting winds.

LAB

19 USING NATURAL RESOURCES

Investigating the Effects of Acid Rain

MATERIALS

- ✓ pH electrode (and adapter cable)
- ✓ CBL
- ✓ TI graphing calculator and black link cable
- ✓ deionized water
- ✓ buffer solution
- ✓ 2–3 samples from different bodies of water
- ✓ 100 mL beakers (5)
- ✓ rinse bottle filled with deionized water
- ✓ large beaker
- ✓ 0.10 M H_2SO_4 in a dropper bottle
- ✓ graph paper (optional)

Wear safety goggles, gloves, and a laboratory apron while performing this experiment.

Sulfuric acid is corrosive. If sulfuric acid gets on your skin or clothing or in your eyes, rinse it immediately with lukewarm water, and alert your teacher. Avoid breathing the acid's fumes.

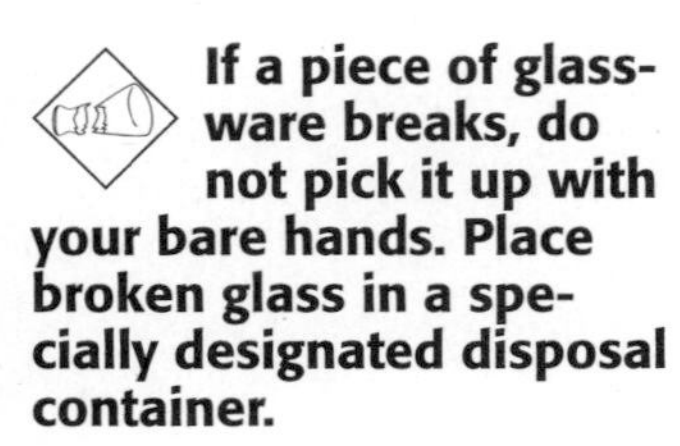

If a piece of glassware breaks, do not pick it up with your bare hands. Place broken glass in a specially designated disposal container.

▸ Objectives

- ***Simulate*** the effects of acid rain by adding drops of acid to several different water samples.
- ***Measure*** the pH of each water sample after adding each drop of acid.
- ***Determine*** which water samples have the greatest buffering capacities.

▸ Identifying the Problem

When fossil fuels are burned, carbon dioxide, water vapor, and other gases are produced. Some of these gases react with water in the air to form acids and eventually fall to Earth as rain. When this acid rain enters rivers, lakes, and other bodies of water, the pH of the water changes and the organisms living in the water are affected.

However, not all bodies of water are affected in the same way. The pH of some bodies of water may change very little, while the pH of others may change a lot. Imagine that you have been asked to do an experiment to determine which bodies of water in your area are the most sensitive to acid rain.

▸ Finding Out More Information

The ability of a solution to resist changes in pH is known as the solution's buffering capacity. A solution's buffering capacity depends on the number and kinds of dissolved ions that are in the solution. Some natural waters have dissolved ions that are basic, such as carbonate ions (CO_3^{2-}) and phosphate ions (PO_4^{3-}). These and other basic ions react with acids. They "buffer" the water so that the pH of the water does not change significantly when acid rain falls. Other natural waters have fewer dissolved ions to buffer the water and are therefore more susceptible to changes in pH.

▸ Coming Up with a Plan

You could simply dump a large quantity of acid into a body of water and then measure the pH of the water. However, doing this would harm the delicate ecosystem that exists there. It would be much better to perform an experiment in a laboratory with several small samples of water from different sources.

If you add the same volume of acid to each water sample, you can compare the pH changes for each sample and determine how well each one is able to resist changes in pH. To make your analysis more complete, you should also analyze samples of water that demonstrate the two extremes of the buffering capacity of water—deionized water and a buffer solution that is made specifically to resist changes in pH.

▶ Performing the Experiment

Preparing for the experiment

1. Prepare a table in your lab report similar to the one shown below to record your data.

pH Data

Drops of acid added	Deionized water	Buffer solution	Sample 1	Sample 2	Sample 3
0					
1					
2					
3					
4					
5					
6					
7					
8					
9					
10					

USE YOUR DATASHEET

When you are done with the experiment, dispose of the waste as instructed by your teacher, and wash your hands thoroughly.

2. Obtain a dropper bottle filled with dilute sulfuric acid. Sulfuric acid can be very corrosive, so be very careful when you are handling it.

Setting up the CBL System

3. Plug the pH electrode into the Channel 1 input of the CBL. Connect the CBL to the graphing calculator by plugging the black link cable into the base of each unit.
4. Turn on both the CBL and the calculator. Press PRGM on the calculator, and select the PHYSCI program.
5. Go to the MAIN MENU, and select SET UP PROBES. Enter “1” as the number of probes. Select pH from the SELECT PROBE menu. Enter “1” as the channel number.
6. Select USE STORED from the CALIBRATION menu.
7. From the MAIN MENU, select COLLECT DATA. Select MONITOR INPUT from the DATA COLLECTION menu. pH readings will be displayed on the calculator.
8. Remove the pH electrode from the storage solution. Use a rinse bottle filled with deionized water to carefully rinse the electrode, catching the rinse water in a large beaker.

Comparing the buffering capacities of deionized water, the buffer solution, and some natural waters

9. Submerge the pH electrode in a 100 mL beaker containing about 50 mL of deionized water. When the pH reading stabilizes, record it in your data table. This is the pH before any sulfuric acid has been added.
10. Place a single drop of acid into the beaker. Swirl the solution gently, being careful not to hit the pH electrode. When the pH reading stabilizes, record it in your data table.
11. Keep repeating step 10 until you have added 10 drops of acid to the sample. (Try to hold the dropper consistently so that the drops are all about the same size.) Record the pH each time in your data table.
12. Remove the pH electrode, and rinse it thoroughly with deionized water. Dispose of both the sample and rinse water as instructed by your teacher.
13. Repeat steps 9–12 for the buffer solution, and then repeat for each sample of water that you have. Record the pH each time in your data table.
14. When you have finished, press "+" on the calculator, put away all of your materials, and clean your work area. Rinse the pH electrode thoroughly with deionized water, and return it to its storage solution, making sure that the cap is on tight. Be sure to wash your hands thoroughly when you are done.

▶ Analyzing Your Results

1. Plot your data on a graph like the one shown below. Plot each sample of water that you tested on the same set of axes, using a different color for each set of data. Be sure to label your graph clearly.

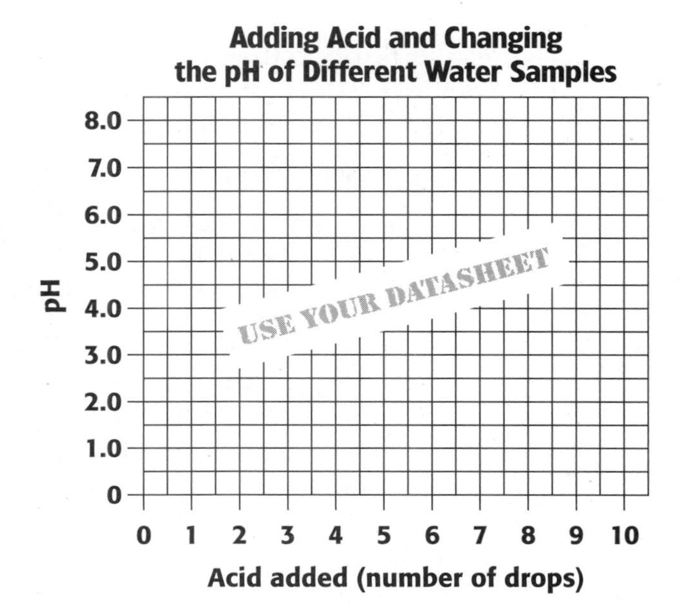

2. Use your graph to compare the buffering capacities of deionized water and the buffer solution.
3. Which natural water sample demonstrated the most resistance to changing pH? Which sample demonstrated the least resistance? Use your data to support your answer.

▶ Reaching Conclusions

4. Why does deionized water have such a low buffering capacity?
5. Explain why some of the water samples you tested have a greater buffering capacity than others.
6. The pH scale is a logarithmic scale. This means that when pH decreases by one unit, the acid concentration increases by 10 times as much. Use this fact to explain why small changes in the pH of a body of water can be critical to the organisms that live there.
7. Reactions within your body are constantly generating acids, such as lactic acid and carbonic acid. You may also eat many foods that are high in citric acid, acetic acid, and other acids. But the pH of your blood always stays between 7.35 and 7.45. Make a hypothesis about the buffering capacity of your blood. Which of the samples you tested is probably most like blood?

▶ Defending Your Conclusions

8. Suppose someone suggests that one way to protect local lakes, streams, and rivers from acid rain is to add a buffer solution to the water. Do you think this is a good idea? Explain why or why not.

▶ Expanding Your Knowledge

1. Test the buffering capacities of several different antacids using a process similar to the one you used in this experiment. Dissolve each antacid in water, then add acid one drop at a time while monitoring the pH. Determine which of the antacids you tested has the greatest buffering capacity.